Endorsements

In the brave new world of AI-generated codebases, this work is an essential companion book for every aspiring programmer. In his characteristic friendly and accessible writing style, Artie quickly introduces you to various ideas, techniques, and principles that are already the mainstay of elite, ultra-productive software engineers. With a great balance between pragmatic advice and theory, the book quickly equips you with the essential street-fighting moves needed to tackle even the most nausea-inducing code complexity horrors. If you want to go from a "coding monkey" to "system architect," then this book is an absolute must-read!

—Jesse Lu, PhD
Engineering Lead, SPINS Photonics Inc

This book is an excellent read, covering in detail often overlooked software development quality practices. It's an invaluable resource if you want to equip yourself with the tools that will help you write sustainable and maintainable code, while fostering a better programming culture.

—Laurynas Kavaliauskas
Senior Principal Architect, Zscaler

This book is filled with actionable guidance and helpful examples. The reframing of certain known ideas and concepts is also timely and instructive. I highly recommend this book to any software professional who wants to level up their knowledge and skills.

—Jeff Doolittle
Senior Software Architect, Trimble

Code Health Guardian offers solid theoretical advice aimed at experienced engineers, all while maintaining a practical focus. Reading this book can help establish or solidify engineering principles that are crucial for maintaining the health of your codebase. The book starts by highlighting that it is meant to be argued with, and the author certainly achieves that goal. Many of the author's views on software engineering mirror my own, while others challenged me to rethink my perspectives. Whether I agreed or not, arguing with the various statements from the book helped me refine my views on various aspects of software engineering.

—Sergey Tselovalnikov
Staff Software Engineer, Canva

Over 20 years in academia, I've observed that Software Engineering majors often focus on Algorithms, Data Structures, and Programming courses; yet becoming a well-rounded software engineer requires much more. Artie Shevchenko, a lecturer at ITMO University, brings a refreshing industry perspective to the classroom. Based on his university course, his book *Code Health Guardian* provides thoughtful, real-world guidance on writing maintainable and efficient code—skills that go far beyond the fundamentals of programming. It bridges the gap between theoretical learning and practical application, offering guidance on how to write software that stands the test of time. Highly recommended!

—Andrew Stankevich, PhD
Associate Professor and Dean, ITMO University
Competitive Programming Coach, ICPC Gold Medalist (2001)

Code Health Guardian

The Old-New Role of a Human Programmer in the AI Era

- Editor: Lana Todorovic
- Proofreader: Cheryl Lenser
- Indexer: Cheryl Lenser
- Cover images: Glebstock/Adobe Stock, Fotomay/Shutterstock

ISBN 978-1-7637717-0-3

For my beloved wife Juliana

Soli Deo Gloria

CONTENTS

FOREWORD

George Fairbanks once noted, "Functionality is localized. Quality is diffuse." That is to say, you can ask where to find the authentication code, but you can't ask where to find the security code: security has to be everywhere in the code. Similar logic applies to performance, portability, observability, and all sorts of other software qualities.

This book focuses on code health, a crucial software quality in software engineering, sometimes referred to as readability, or software maintainability. If we are writing short-term throwaway code for ourselves, we can cut corners. But professionally, most of our code is for consumption by other humans, as well as ourselves in the future. What I've found, unfortunately, is that for any *quality* attribute we care about, we need to be diligent to avoid sliding down the slippery slope.

As we enter the AI Era, these lessons are more pressing than ever. At present, with LLM-powered AI coding assistants, we're accelerating the production of code, especially in the hands of junior engineers. I have seen no evidence, nor do I have any reason to believe, that this form of AI is going to be a net positive on code health. As Ted Chiang wrote in *The New Yorker*,[1] AI is a lossy compression over the input corpus: we ought to generally expect that on a chunk-by-chunk basis, AI agents produce code at roughly the average quality of the training data. For foundational models, that's likely to be the publicly available code on the Internet. While there is some good work being done on fine-tuning and localizing such models, it's hard to envision an outcome where a lossy compression of the training data results in an average quality improvement.

---

[1] https://www.newyorker.com/tech/annals-of-technology/chatgpt-is-a-blurry-jpeg-of-the-web

I for one don't feel reassured by the idea of accelerating our production of (best case) average quality code. Don't get me wrong, I love having these tools available, and I think they offer substantial advantages for many developers. I rely on them myself, especially when I'm coding in a language that I'm not deeply familiar with. But unlike context-aware autocomplete tools in our IDE in the 2010s, these agents eagerly *try* to produce increasingly bigger suggestions. Without humans in the loop actively guarding for quality, I fear we are going to go off the rails.

In this book, Artie has assembled a truly excellent collection of practical advice for reducing those risks. In clear (but detailed) language, this book lays out the case for better human involvement in monitoring for code health, as well as pragmatic guidance for getting started. While it's difficult (if not impossible) to teach *taste*, the lessons in this book are great at conveying practices that will on average lead to better software design and code health.

I am hopeful that this book will become an essential reference for software engineers, giving solid basic advice on day-to-day topics such as interface design, contracts, coupling, naming, code review, and documentation, just to name a few.

For the junior engineers out there: take this book as a chance to think about *how* you practice your craft, rather than only the functionality you can assemble.

For the mid-career engineers: use this book as a chance to revisit ideas that you saw years ago but didn't understand the need for. It's a great refresher, and especially potent if you tackle this at the same time as you look at adopting new AI tools.

And for the engineering leaders out there: a copy of this book is a one-time cost, probably on par with the monthly cost of AI developer tools for a single user for a month or two. Can you afford not to have a copy handy for each of your development teams? How much will you regret it in a few years when the AI-powered bump on coding speed is eaten away by the messes produced by this year's AI agents?

I'm flattered to have been cited by Artie several times through the text, but I'm just as excited for all of the ideas he has drawn from other thinkers and repackaged here. I hope you all enjoy it as much as I have.

Titus Winters
New York
October 2024

PREFACE

The subtitle of this book, "The Old–New Role of a Human Programmer in the AI Era," may make you believe this book is about how to collaborate with AI agents. But it's not nearly about that, for two very simple reasons. First, I don't believe it to be the crux of the problem. Basically, you can be among the first to master a certain tool or prompting technique, but if it truly makes the difference, everyone else will catch up in a couple of months anyway. This reminds me of my mom trying to teach my son English while visiting us in Australia (she is an English teacher in Russia). The thing is, in just a few months, he'll be able to teach *her* English, given that now he goes to school and has English-speaking friends. In the same vein, over the coming years, we'll surely learn mostly organically how to collaborate with AI agents.[2] The second reason is that we're not there yet, and anything we know today about guiding AI to do what we want is likely to become suboptimal or even obsolete with the release of the next breakthrough AI model. So instead, let's focus on what I believe is going to be at the very center of the old–new role of human software engineers in the upcoming AI era, that is, how to keep our codebases reasonably simple. To fully explain the reasoning behind this statement, let me start from afar.

In 2017, I met a guy from Google X, Google's secretive moonshot factory, who also held a PhD from Stanford. Although he never shared anything about his projects, he once briefly mentioned he was saving Google millions of dollars per month. To my surprise, he also habitually referred to himself as a coding monkey. At first, I criticized him for doing so, but very soon I found out that he wasn't alone—in fact, the question of whether software engineering is worthy of being called engineering is as old as programming itself. In his classic book

---

[2] Similar to how many of us have been learning to interact with GitHub Copilot recently. `https://docs.github.com/en/copilot`

The Mythical Man-Month, Fred Brooks wrote, "Parnas[3] reproves me for writing about software engineering at all. He contrasts the software discipline with electrical engineering and feels it is a presumption to call what we do engineering" [Brooks95]. To take it even further, recently, one of Canada's provinces made it *illegal* to call programming "software engineering"![4] Although at first it may sound ridiculous, it is in fact quite logical. Most of the code in production, both today and 30 years ago, is a weird mix of what we call best practices combined with authors' own ideas and accumulated experience. It's just a fact. And we must admit—it doesn't sound like a solid science-grounded engineering discipline.

Not only is there a huge gap between computer science and software engineering, but we're also not on a trajectory to narrow it [Barr18]. There is a simple explanation for that: problems from mainstream software engineering look neither interesting nor promising enough for those in academia. On the other side of the fence, we programmers don't see much value in scientific research as the vast majority of it is impractical, and any rigorous analysis of most real-world problems seems to be unfeasible. Thus, a growing gap between programming and computer science has been a rather stable status quo for decades.

Yet today, AI is here to disrupt our industry and challenge this status quo. The rise of AI has already begun elevating the role of science across our field, but it will soon do so at scale, elevating the role of science even for typical software engineering jobs. It's highly likely AI agents will become the best coding monkeys on the market, but what will be much harder for AI to master is managing code complexity, or as I call it in this book, keeping code healthy. Indeed, that's the most intellectually challenging problem in software engineering, so the role of the Code Health Guardian currently seems to be

---

[3] David Lorge Parnas is a Canadian early pioneer of software engineering, who developed the concept of information hiding in modular programming, which is an important element of object-oriented programming today.
`https://en.m.wikipedia.org/wiki/David_Parnas`

[4] `https://www.cbc.ca/news/canada/calgary/tech-companies-alberta-premier-software-engineer-title-1.6617742`

reserved for humans. Yet, in that role, understanding the theory behind code complexity is almost as important as your practical experience. That's already the case today, and it will become even more so tomorrow.

Many years ago, a friend of mine asked me why I care so much about theory instead of focusing solely on practice. Being such a good student at the time, I responded with a counter question: "But then, how would you deal with any new situation you encounter if your practice is not rooted in theory?" I don't think he was able to answer that, and that's how our short conversation ended. However, that's not the full story. It was a genuine question from his side as he's a man of practice. In particular, driving a car was a part of his job at the time, and he had really mastered it, so that when driving, it seemed he could literally sense the boundaries of his car. But it was him who was driving when I got into the scariest car accident of my life, just a few months after that conversation. And he wasn't speeding or anything like that. It was a rainy night, and we were driving on a highway. I remember those 20 seconds as if it were yesterday. When the left wheels of our car hit a rut full of water, at first, it looked like everything was still under control. The car began to slide and sway, but it was a barely noticeable movement. Yet the amplitude increased with every next sway, and in just several seconds, we were sliding along the highway at full speed, spinning and crashing into the guardrail and other cars. As you may guess, the problem was that my friend released the gas pedal and tried to level the car with the steering wheel. Despite him being one of the most experienced drivers I've ever known, he made a mistake that could have cost us our lives. And this happened simply because he was too focused on practice and ignored theory. Lack of theoretical knowledge in our field may not have such drastic immediate consequences, but it definitely makes it much harder for long-running projects to survive and for us to properly do the job of a Code Health Guardian, that is to keep our codebases reasonably simple.

There are so many coding ideas floating around that it's almost impossible to navigate them. It's so easy to accept and mimic some experts' practices, or something from your own past experience, or even an established industry practice, just to realize months later that it turned out to be counterproductive.

It happens all the time! Therefore, this book has three primary objectives. It aims to:

1. Explain the theory behind code complexity
2. Use that theory to evaluate a bunch of industry's best practices
3. Train you to apply it to any code complexity-related problem that you may face in the future

When discussing coding practices, this book is slightly more scientific than you may be used to. However, it is exactly this kind of approach that empowers us to get the insights that we wouldn't get otherwise.

It's important to find the right balance between theory and practice in everything we do, and I hope this book makes its modest contribution to closing the aforementioned gap between software engineering and computer science. But most of all, I hope it will make your code simpler, your software less fragile, your coding less stressful, and ultimately help you qualify for the old–new role of human software engineers in the upcoming AI era—the role of a *Code Health Guardian*.

Sydney, Australia
September 2024

How to use this book

This book tells the story of a healthy codebase—a codebase that is reasonably simple. After the introductory chapter, the book is divided into two parts: Part I: "The Code" (Chapters 2–7) and Part II: "The Key Code-Related Processes" (Chapters 8–10).

Chapter 1: "Introduction" demonstrates how vital code health is for productive development and highlights the importance of always striving for the pragmatic balance between short- and long-term productivity. Next, Chapter 2 introduces a code complexity model that is used throughout the book. It provides just enough theory to understand the relative value of the principles and techniques discussed, so that you can prioritize them and maybe even focus on the most important ones on your first read.

The order of chapters 3–10 generally doesn't matter, so feel free to read them in an order that's most meaningful to you. Chapter 3: "Design by Contract" arguably covers the most fundamental software engineering technique. It focuses on tests and assertions, given that these contract-enforcement mechanisms are widely applicable to both strongly and loosely typed languages. Chapter 4: "Interface Simplicity" introduces the concept of interface depth; analyzes the trade-offs of small, deep, and shallow modules; and covers several interface simplification techniques. But most importantly, it sets the tone for the rest of the book, demonstrating the importance of the interface-first mindset. Chapter 5: "Independence Quest" offers a fresh perspective on the software design principles that you're likely very familiar with and don't expect to learn anything new about them. Among other things, it illustrates how the theory from Chapter 2: "The Code Complexity Model" helps evaluate the relative importance of different best practices.

There are several chapters that I expect to be the most influential, as, from my experience, not that many developers practice what is being discussed there, probably assuming it's unrealistic. Chapter 6: "Semi-functional Programming" and Chapter 8: "Trade-off Analysis and Design Docs" each in its own way challenge the industry's status quo and stimulate you to try to do things differently. Both of these chapters offer pragmatic approaches that give you 80% of value for 20% of effort. Chapter 6 specifically covers immutability practices that deserve much wider adoption. Part I ends with Chapter 7: "Naming: The Obscurity Champion." Besides the subject matter discussed there, this chapter is full of real-life stories, insights, and is just fun to read.

Chapter 8: "Trade-off Analysis and Design Docs" opens Part II: "The Key Code-Related Processes." Without the second part, this book would tell only half the story of a healthy codebase, as code health starts outside of code. In this book, I apply the trade-off analysis technique from Chapter 8 to two practical problems—choosing between deep, shallow, and small modules (see Chapter 4: "Interface Simplicity") and between the build and buy options (see Chapter 9: "Build vs. Buy"). Understanding these two applications would be a great practice for those who are new to the formal trade-off analysis techniques, but, most importantly, these two chapters make it crystal clear what I actually mean by formal trade-off analysis in Chapter 8.

Chapter 9: "Build vs. Buy" covers a topic that is likely to be of interest only to senior developers. However, if you're skipping any chapter, it's worth reading the "Summary" section, which every chapter of this book ends with.

Finally, Chapter 10: "Code Reviews" covers the central code-related process. Code reviews are where the story of a healthy codebase is being written, so getting them right is tremendously important. Learning from Google's wisdom derived from millions of code reviews done by hundreds of thousands of developers over two decades looks like a smart move. I advise everyone to read that chapter.

Code examples in this book are written in Java. They're optimized for brevity, just to illustrate a specific point, so they may not follow all the best practices and even ignore some of the recommendations from other chapters of the book. Please don't consider the provided snippets to be production-ready code.

I'm convinced that books must be studied and argued with, not just be read. While it's very individual which sentences or paragraphs to meditate on, I think there are some common breakpoints in this book that are worth stopping at for everyone. These are marked as

---

II Pause and Think

Please treat such breakpoints as gentle nudges to check whether you're just consuming content or are truly studying and reflecting on what you're reading. For the very same reason, this book has many footnotes—also, treat most of them as nudges to argue with the book and meditate on the content.

Furthermore, numerous external links and references are provided. Treat them as optional further reading—feel free to ignore them unless you're particularly interested in delving deeper into the topic.

The best way to communicate with me about this book is to send an email to the following address:

`code-health-guardian@googlegroups.com`

If you'd like to see what others are saying and participate in discussions, you can join the `code-health-guardian` Google Group. If for some reason that Google Group disappears in the future, you can find updated instructions on how to communicate about the book at `https://codehealthguardian.com`. On the same website you can also find information on how to report typos.

Acknowledgments

First of all, I carry a heavy debt of gratitude to my wife, Juliana, and my children, Daniel, Kathy, and Eva. Juliana, thank you, dear, not only for enduring my three-year-long obsession with writing this book but, perhaps even more importantly, for supporting my decision to leave Google and return home to Russia to focus on the startup and university lectures. Your willingness to embrace my ideas and ideals, even if it meant leaving the comfort of life in the United States, is something I deeply appreciate. When we got married, you promised to follow me wherever I go, and you've kept that promise. I'm truly thankful and proud of you.

There are three key events in my life that made this book possible. The first one is John Ousterhout's book presentation at the Sunnyvale Google office in 2018. I still believe *A Philosophy of Software Design* [Ousterhout18] is one of the best books for developers, and together with *The Pragmatic Programmer* [HT19], these two books inspired my university classes, which ultimately led to the creation of this book. Dr. Ousterhout, thank you for your work, for the `software-design-book` Google Group, for responding to all my emails over the course of the last five years, and for your best wishes for the *Code Health Guardian* book project.

The second key event is when I started the Good Code Reviewer classes internally at Google. I believe it's teaching those classes that gave me the necessary experience and courage to later teach at the university. Code review classes (as well as the last chapter of this book) were mostly based on Google Engineering Practices recently made public.[5] Although your name isn't explicitly mentioned there, Max Kanat-Alexander, I know you're the author of most of it. It's your wisdom that ultimately made those classes possible, and I want to express my gratitude for your support in that initiative.

Finally, the most important event is starting the Productive Software Engineering classes at ITMO University. Andrew Stankevich, thank you for being open to something new and giving me a chance (there are details in Chapter 7). This book was born out of those classes, with me iterating on their content for four years now. But of course, it's not just me iterating—I want to express endless thanks to my students for their feedback and counterarguments. Truly, there's a reason why ITMO has more ACM Programming World Cup[6] titles than any other university in the world. Engaging in discussions with you was often a challenge, but I'm grateful for taking it on.

Titus Winters, your support and advice at the early stages of writing the book were invaluable. I know it ended up being a slightly different book than originally discussed, but that's what time and the release of ChatGPT brought with them. I'm especially thankful that you agreed to write a foreword to this book!

Leo Osang, thank you for a gentle nudge in early 2024 to finish this work, which was collecting dust on the shelf for most of 2023.

---

[5] `https://google.github.io/eng-practices/`

[6] ACM ICPC (International Collegiate Programming Contest). `https://icpc.global/`

Jesse Lu, I believe you were the earliest collaborator on the content when I was drafting the "Semi-functional Programming" chapter. Thank you for having so much patience with a first-time author.

This book has been tremendously improved by the suggestions and assistance of many reviewers. I would like to thank Adam Barr, Amy Wang, Ankit Ashri, Alex Shestakov, Melissa Ermishina, Tin Pavlinic, Erika Shefer, Sergey Kazakov, Viktar Sharepa, Aleh Boitsau, Egor Nazarov, Taras Skazhenik, Denis Vorkozhokov, Igor Podtsepko, Ilia Zaitsev, Rahim Nakimov, Nikita Kozhukharov, Laurynas Kavaliauskas, Alex Gemberg, and Kim Heusel, who reviewed the book at different stages, provided feedback, and made suggestions that helped make this book better.

Lana Todorovic, it truly was an amazing experience editing the book with you. Your meticulous work in fixing my English and making it easily readable was invaluable. I appreciate your dedication, attention to detail, and especially your humbleness.

Cheryl Lenser, thank you for your incredible work as both indexer and proofreader. Your sharp eye, experience, and precision ensured every detail was perfect, and your contributions have truly improved this book.

There was one traditional publisher that I shared the book proposal with, and that was O'Reilly. I want to thank you Louise Corrigan for your time—even though you were hesitant about publishing it, our relatively short interaction has definitely helped me with the publishing process.

Finally, I want to express my gratitude to my manager and teammates at Canva for having my back during the rollout of the ambitious project we're working on right now. I recognize that taking a couple of months off during the final stages of working on this book was less than ideal. Justin Velluppillai, thank you for all your support along the way and for approving my countless leave requests!

Chapter 1

INTRODUCTION

When it comes to software complexity, I will let the legends set the stage for us. In the 20th-anniversary edition of his classic book *The Mythical Man-Month*, Fred Brooks says:

> *The distinctive concerns of software engineering are today exactly those set forth [twenty years ago]: ... how to maintain intellectual control over complexity in large doses.* [Brooks95]

Furthermore, Tony Hoare, although speaking about software complexity somewhat categorically,[7] perfectly conveys the very crux of the problem in his 1980 Turing Award Lecture:

> *There are two ways of constructing a software design: one way is to make it so simple that there are obviously no deficiencies and the other way is to make it so complicated that there are no obvious deficiencies.* [Ashenhurst87]

Finally, the last quote comes from John Ousterhout, whose book *A Philosophy of Software Design* highly influenced this work:

> *The greatest limitation in writing software is our ability to understand the systems we are creating.* [Ousterhout18]

---

[7] Section 1.2 probably explains why.

Pause and Think

I have no doubt that these quotes resonate well with every experienced programmer. Too often does complexity win over teams, but it shouldn't be that way. At the very least, we want a draw!

Code complexity is one of the biggest—if not the biggest—challenges in software engineering. To start with, codebase of any nontrivial product is inevitably somewhat complex, which is mainly due to the complex nature of business requirements: if users want a program to do 30 different things, there will be a complexity cost to pay, regardless of the language and technologies used. To make it work fast, there may be even more unavoidable[8] complexity to add, and the same is true for other non-functional requirements: security, reliability, privacy, and so forth. Complexity is often a necessary trade-off. That's why, pragmatically speaking, we aim at having a healthy codebase, not a simple one. Sometimes, I would still use a well-established "simple code" phrase in this book, but keep in mind that, strictly speaking, there's not much production code in the world that is truly simple in the full sense of that word.

That said, code is not responsible for *all* the complexity in software projects: it may be a complex UI, chaos in the database, burdensome development methodologies, organizational complexity, complex tooling, and so forth. However, the codebase, if complex, typically dominates every other problem.[9] So, in this book, *code complexity* is implied whenever the term complexity is used without any clarification.

---

[8] Not necessarily *essential complexity*, as that term is commonly understood. It may be unavoidable in one technology stack and absolutely unnecessary in another.
[9] It doesn't have to be execution code, which first comes to mind, but configurational code as well. Configuration can be complex in itself, or it can stimulate complexity growth indirectly (e.g., the way bad database schemas provoke data chaos).

1.1. Productivity implications of complexity

It is not an overstatement to say that a complex codebase poses an existential threat to any long-term project. The first reason for that is that it "sucks the life out of developers," as per Ray Ozzie, former CTO of Microsoft [Martin08]. That by itself is so serious that we could have stopped here and proceeded with Chapter 2. I know I may be biased, as I'm a software engineer myself, but even if you're not a developer and don't care much about their wellbeing, there are still good reasons for you to care about code health. Apart from exhaustion and increased turnover, there are at least two other major productivity problems that code complexity brings with it:

1. **Complexity makes the product buggy.** You can never be sure that a complicated piece of code is bug free. Bugs tend to hide for long periods of time, sometimes even compensating for each other, just the way two erroneous minuses in math do so. But even if there are no bugs in complicated code yet, it's not for long—just give it enough time and iterations. To make things worse, when fixing a bug in complicated code, it's super easy to introduce another one, which is the first vicious productivity regression cycle in Figure 1.1.
2. **Complexity slows you down.** Obviously, in a complicated codebase, it takes more time than it should to accomplish things, which is often an even bigger problem for a product (and product managers) than bugs. And unfortunately, "adding manpower to a late project just makes it later."[10] [Brooks95]

So, here are three main productivity problems coming from complex code: developers are exhausted, product is buggy, and development speed is low.

---

[10] Depending on how complicated your code is, you may be able to improve the development speed short-term by adding the right developers with the right experience. However, growing the team beyond a single-digit number of programmers in the long-term would also accelerate the growth of complexity, which will ultimately slow you down, no matter how much you increase the project's budget.

Once accumulated, these problems quickly shift any project into maintenance mode.

As you may have already noticed, these three problems are interconnected (adding more arrows to Figure 1.1):

- Bug fixing slows you down.
- "Quick fixes" introduce more complexity.
- Exhaustion doesn't help with the development speed and incident management (and the quality of the product overall).

Pause and Think

Last but not least, chaos in code is contagious. Not only are people afraid of refactoring the code they don't understand, but also having a broken window causes other windows to be broken soon, as it is perfectly depicted in the "Software Entropy" chapter of *The Pragmatic Programmer* [HT19]. The feeling that nobody cares about a little bit of additional complexity is appealing and contagious. That's yet another way how code complexity multiplies itself, introducing the final vicious cycle to our productivity regression diagram (see Figure 1.1). I think it already looks bad enough for us to get scared, wrap up this section, and move on.

1.2. Complexity: The silent killer of software projects

The productivity implications of a complicated codebase are scary, but what makes it such a dangerous enemy is that *complexity may look pretty innocent until it's too late.* Ignoring complexity is like slowly accelerating on a wet road—you're going increasingly faster, and everything seems to be okay and even fun until you start losing traction. Way too often, teams cope with growing complexity for a long time, barely noticing how complicated it already is. Progressively, we just get used to the fact that we need to spend a bit more time exploring the codebase to accomplish a task at hand,[11] until one day we wake up and realize that our codebase is super confusing. What makes this scenario likely is the fact that software engineers are smart—we can digest quite a lot of complexity before even noticing it (see Figure 1.2). But by then, it may be too late—productivity problems start growing too fast when one is operating near the limits of their cognitive abilities.

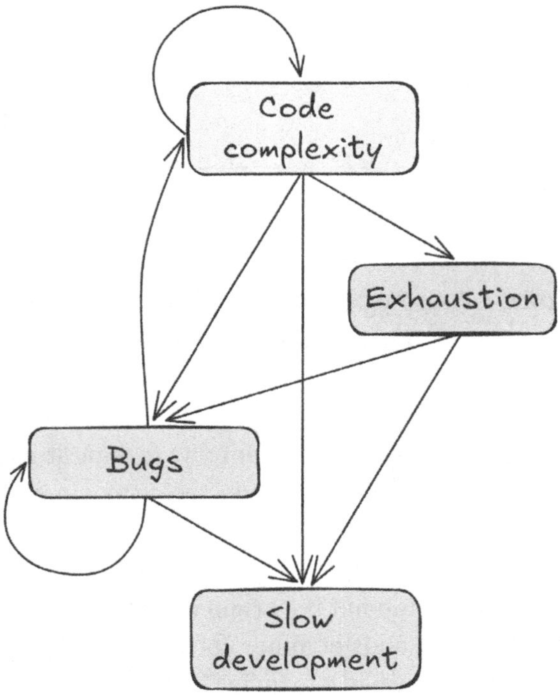

Figure 1.1. Productivity regression in complicated codebases.

[11] And not break something else.

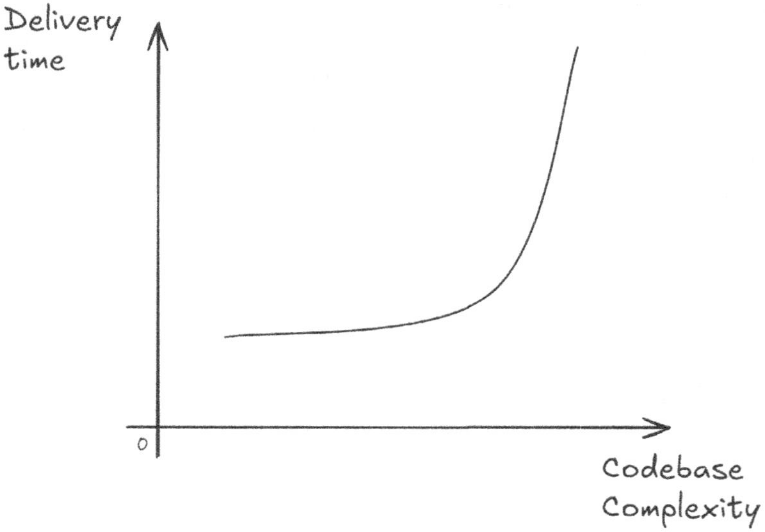

Figure 1.2. Growing complexity can be manageable and even unnoticeable for quite some time.

Finally, the worst thing about the codebase complexity is that there is almost no way back. Complexity grows incrementally, making it extremely hard to reverse the damage. At some point, you may decide to win it back—to refactor your code into something reasonably simple, but after multiple weeks of dedicated refactoring efforts, there would likely be very little progress just because it's impossible to pinpoint the origin of all that complexity. And there is no single origin. It's thousands of minor complexities that accumulated and interconnected over time. We had a team in Google Maps that was moving fast until they couldn't, so they had to take a break and spend over a year just refactoring their code. This is not something every company can even afford.

You can think about accumulating complexity as driving on a highway section without any exits. You move fast, accumulating complexity without even noticing the speed you're driving at. Yet, deciding to pull over and back up to take the previous exit would be risky, and you'll definitely be surprised how far you've gone. Refactoring code to make it simpler, very similar to backing up on a highway, is slow and painful, and starting from scratch may be the only alternative.[12]

Avoid putting yourself in a situation where there are no good solutions. It's important to remember that code complexity often goes unnoticed and is extremely hard to undo. Be constantly killing complexity, or it will be killing you down the road—both your mood and your project.

1.3. *-term productivity spectrum

Moving fast, ignoring the complexity issues, is programming *tactically*. Even though it may be okay when prototyping or working on a tiny project,[13] continuous tactical development for a typical project inevitably leads to a dead end. Even short-term, it's usually not the best choice to cut corners for the sake of fast delivery. In the words of Max Kanat-Alexander:

> *Some people believe that writing simple code takes more time than quickly writing something that "does the job." There is no data of which I am aware that validates this idea. Serious software development nearly always has long timelines—weeks or months at the shortest. When you add complexity into your program, you're slowing yourself down tomorrow.* [Kanat-Alexander12]

In contrast, programming *strategically* is about maintaining a healthy codebase along the way. There is a widespread notion of a 10x-developer,[14] but there is also a common misconception about what makes them 10x more productive than others. Realistically, you can hardly imagine one professional well-motivated programmer doing an order of magnitude more than another professional well-motivated programmer. Instead, what makes the difference

---

12 Starting from scratch forces you to learn all the subtleties of the domain all over again, which can be slow and painful, too.
13 That's a thin-ice situation, though, as big projects often start small.
14 A developer that is supposedly as productive as 10 typical developers combined.

is that the 10x-developer empowers others so that nine additional developers' worth of work gets done. And that's exactly what strategic programming is—maintaining a healthy codebase to enable others to work with that code productively.

This challenges another common misconception that only feature work is product focused, while the same cannot be said for the busy work of keeping the codebase healthy. In fact, both types of work, if done properly, are product focused. As an engineer, there is not much you can do to find the product–market fit or develop the right marketing strategy, but there are two things *only you* can do to help your company succeed and make your users happy:

- **Short-term productivity:** Deliver high-quality software right now.
- **Long-term productivity:** Write healthy code to deliver high-quality software a year from now.

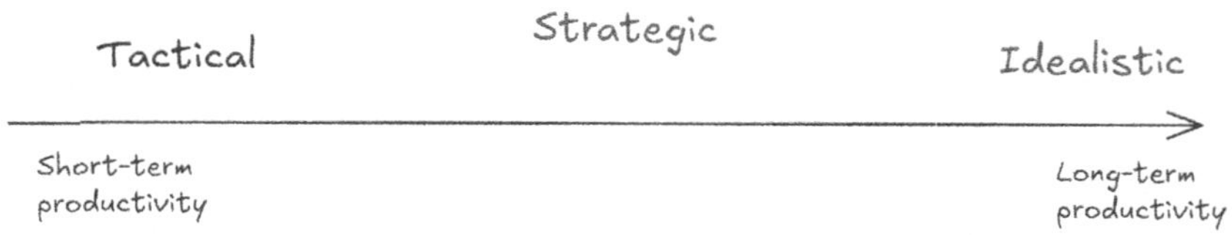

Figure 1.3. *-term productivity spectrum.

Every well-motivated software engineer falls somewhere on the *-term productivity spectrum (see Figure 1.3). Tactical programmers, instead of finding the right balance, focus on short-term productivity, and theorists and idealists focus on the long-term. It's not to say that you should never do that. Sometimes you may deliberately choose to focus on short-term goals for weeks or even months, for users' sake. And sometimes you may choose to focus solely on long-term goals, again for users' sake. But that should definitely not be your default mode as a developer. Strategic programmers, programmers who truly care about the product, constantly balance between short- and long-term productivity. They care about their users being happy both today and in a year from now.

1.4. Embracing simplicity: The hallmark of software engineering

When I was a student, one of our university teachers began his lectures this way: "You may think you already know how to write software. But you don't. Some of you know how to solve extremely complex coding problems and write tons of working code in five hours,[15] but that doesn't make you a software engineer." I still remember the silence in the auditorium after such an intro. It was my third university year. Some of the students in the room were ACM ICPC[16] champions already, and most of us had been coding for five to ten years at that point. Some even had a little bit of experience in big tech or an offer at hand. But he was right—we were not software engineers yet. As Brian Randell perfectly defined software engineering, it is "the multi-person development of multi-version programs" [Parnas11]. While programming happens in a simple, purely technological world, software engineering happens in a world with two additional dimensions: time and people. To better feel the weight of this change, just imagine a creature that is used to living in a 1D space. It may be an expert in living on a line, but if in the blink of an eye it's transferred into the 3D world we're all familiar with, it would be pretty helpless, right? It's not to say that all of its previous knowledge and skills would be in vain, but they would definitely have a very limited applicability in the world with the two additional dimensions. That's programmers on their first encounter with software engineering.

> While programming happens in a simple, purely technological world, software engineering happens in a world with two additional dimensions: time and people.

[15] Referring to ACM ICPC (International Collegiate Programming Contest) format.
[16] `https://icpc.global/`

One of the biggest differences between programming and software engineering are drastically different consequences of code complexity. That's where the fun definition of software engineering famous at Google is coming from: "It's programming if 'clever'[17] is a compliment, but it's software engineering if 'clever' is an accusation."[18] Simplicity plays such a huge role in the world of software engineering precisely because of the two new dimensions: time and people. An average program needs to be supported and improved over a long period of time, often by many different people. What looks understandable for an author today may confuse them six months from now. And what about others? To make things worse, what if the author leaves the company? A piece of complexity that may be okay and even cool in the world of programming is likely to be destructive and unacceptable in the world of software engineering.

Thus, you can say that attitude toward complexity distinguishes software engineers from mere programmers. It is possible to be a senior programmer and a junior software engineer at the same time. As Martin Fowler puts it, "Any fool can write code that a computer can understand. Good programmers write code that humans can understand" [Fowler99]. By the way, Fowler says "good programmers," not "good software engineers," and henceforward in this book, I'll be using these terms interchangeably as well. Just keep in mind that, strictly speaking, we won't discuss programming in this book—it's 100% about software engineering.

> A piece of complexity that may be okay and even cool in the world of programming is likely to be destructive and unacceptable in the world of software engineering.

[17] Referring to "clever" complicated solutions.

[18] Titus Winters, *Software Engineering at Google: Lessons Learned from Programming Over Time* [WMW20].

If a program's lifetime is several days, and there are just a couple of you working on it, you're doing programming, and you may be totally fine with a purely tactical approach. But any long-running project by definition belongs to the software engineering space and requires a more strategic approach of balancing between short- and long-term productivity, with code health definitely being one of your top priorities. That's what it means to be a software engineer, not merely a programmer.

Summary

While there are many aspects of a software project that can make it complex, source code is the main complexity contributor, which typically dominates everything else. There are three interconnected productivity problems that a complex codebase introduces: developer exhaustion, bugs, and slow development. Code complexity is highly contagious and reinforces itself in multiple ways (see Figure 1.1). Furthermore, it typically grows incrementally as thousands of small complexities accumulate and interconnect over time, which makes it extremely difficult to reverse the damage. Unlike programming, software engineering requires a strategic approach to complexity. Normally, you should be somewhere in the middle on the *-term productivity spectrum (see Figure 1.3). As that's what most long-running projects require from us—to constantly balance between short- and long-term productivity, with code health being one of our top priorities.

Part I

The Code

Chapter 2

THE CODE COMPLEXITY MODEL

When somebody refers to a piece of code as being complex, what they typically mean is that it has poor readability. We often think of code complexity as if it were a single problem, and that's also how it was presented in the previous chapter. However, in practice, there are three distinct problems at play here: high cognitive load, change amplifications, and unknown unknowns. Being able to tell them apart is very useful as they infrequently go hand in hand and greatly differ in severity. If you're familiar with the code complexity model from *A Philosophy of Software Design* [Ousterhout18] by Dr. Ousterhout, then Section 2.1 will also look very familiar to you. But note that what Dr. Ousterhout calls "complexity *symptoms*," I call "complexity *problems*."

In addition to the three complexity problems, this chapter also introduces the seven underlying causes of complexity:

- Obscurity
- Duplication
- Unfamiliarity
- Too many dependencies
- Unstable dependencies
- Untrue interfaces
- Complex interfaces

Chapter 2 sets the *foundation* for the rest of the book. It introduces the code complexity model that will be assumed and extensively used in the chapters that follow.

Finally, in addition to setting the foundation, Chapter 2 also provides a *book overview*. It outlines the story of a healthy codebase, providing references to different parts of the book. So you can use it as a map[19] to navigate to the topics of most interest to you.

2.1. Complexity problems prioritized

Let's begin with the three practical code complexity problems, in the order of their increasing severity.

Change amplification. The first problem is when a seemingly simple change requires code modifications in many different places, which is often referred to as a lack of orthogonality. For example, let's say your code uses Acme API: in one place, you make a call to `api.acme.com/price/current`; somewhere else you make a call to `api.acme.com/price/historical`; and so on. If all the Acme API URLs are hardcoded, and you need to change the base URL from `https://api.acme.com/` to something else, you would obviously have to make changes in multiple places. In contrast, if you have a single `getAcmeApiBaseUrl()` function and use it consistently across the codebase, it would be enough to change the body of that function only:

```
String getAcmeApiBaseUrl() {
  return "https://api.acme.com/";
}
```

Change amplifications may be time-consuming, but thanks to clipboard and find-and-replace tools, it's often a bigger problem for code reviewers than for the author of the change.

Cognitive load. The second complexity problem is when it's hard to understand how the code works, and that's the problem we typically think of when referring to a piece of code as being complex. For example, a piece of code

---

[19] The "How to use this book" section provides the first such map.

may be presenting too much information at once, require multiple steps of complex reasoning, or use concepts that are novel or too abstract. All that increases the cognitive load on our brains and may impair our ability to understand, debug, or modify code. That's the case with red–black tree or hash table implementations, for instance. Every implementation of those algorithms imposes a high cognitive load on the reader, because of their inherent complexity. High cognitive load introduces the risks of wrong interpretation, sometimes even raising the bar for who can actually work with that code.

Unknown unknowns. The third complexity problem is that you may not have all the facts you need to make a code change. You can spend hours reading the code, you can be super smart, and even ask your teammates for help, but still miss something. In fact, in the base URL example above, change amplification is not necessarily the only problem. We've made an implicit assumption that you can search for `api.acme.com` across the codebase and modify its every occurrence. But what if one of the calls looks like this:

```
String.format(
    "https://%s.%s/%s", apiSubDomain, acmeDomainName, query);
```

And it's simply because somebody anticipated the need for the base URL change in the future. They meant it for good but didn't have capacity to consolidate it with the rest of the code (sounds familiar, right?). Or, there may be an unfortunate line break:

```
"https://api.acme." +
       "com/price/historical"
```

There are plenty of reasons why these API calls might look different, which makes it virtually impossible to be confident that you've found all of them with text search and regular expressions. If not centralized, the only way to fully de-risk it is to skim through all the code that can theoretically make an Acme API call. Even if possible, it is tedious, exhausting, and therefore prone to errors.

The problem of unknown unknowns often *involves* high cognitive load, but it's more than that. If there are no unknown unknowns, you may just need to spend more time to understand the code, and that's it. In extreme cases, you might not be able to understand it yourself, but for sure, you can find somebody who can. With unknown unknowns, it's a whole different story. You may be lucky and not miss any important details, or you may not be so lucky. It doesn't help much if you're super smart, experienced, or spend hours on it—you can still miss something. In the presence of unknown unknowns, you may need to be lucky. That's what makes it the worst of the three complexity problems.

> In the presence of unknown unknowns, you may need to be lucky. That's what makes it the worst of the three complexity problems.

This classification helps us clearly see how complexity problems differ in severity and prioritize them accordingly. Figure 2.1 is a slight modification of Figure 1.1—the code complexity box there is replaced with the three more specific complexity problems. Unless the amplification is huge, change amplification is a lower-priority problem. It would slow you down a bit but wouldn't cause any other productivity issues, unlike cognitive load and unknown unknowns, which both cause emotional exhaustion and bugs, and reinforce themselves.

Pause and Think

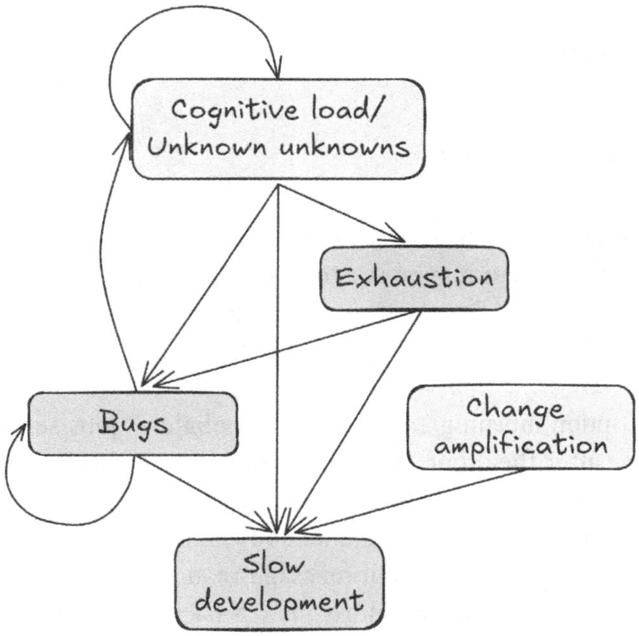

Figure 2.1. Complexity problems and their productivity implications.

You can definitely use a single term, such as "readability" or "complexity," but distinguishing between these three complexity problems empowers you to more objectively assess the value of different best practices, based on the severity of the complexity problems they address. For example, in Chapter 5: "Independence Quest," we'll see how a DRY principle violation, generally being a high-priority problem, can be easily transformed into a lower-priority one.

Differentiating between the three complexity problems helps us be more pragmatic about code complexity. I'm sure you've seen developers being religious about some of their rules and practices, justifying it by a noble goal of simplifying code (me certainly being one of them in the past). While sometimes you may have felt there's something wrong with that justification, you probably couldn't explain what. That's where this code complexity model shines—it provides us with a mental framework to reason about the severity of a specific code health problem. Thus, it enables us to develop a well-considered stance on the issue of code complexity—taking it seriously, yet avoiding the pitfalls of clean code fanaticism.

2.2. Chesterton's fence

Thanks to Google's mature processes, reliable infrastructure, blameless postmortem culture, and maybe my teams' luck, I didn't see many incidents during my four years at Google. That makes me clearly remember those few that we actually had. Let me share one of them.

On a sunny London morning, most of our daily batch[20] jobs suddenly started alerting us one after the other. It was a storm of alerts. None of the jobs could start because of yesterday's jobs that were still running for some reason. It didn't take us long to figure out that somebody had refactored one of our batch job configuration files. The commit message read something like, "You don't really need all of this stuff anymore." It's a tricky situation that I'm sure everyone can relate to—when you want to make a change but don't fully understand the reasons behind the existing code. What do you do then?

The Chesterton's fence principle is named after the English writer and philosopher G. K. Chesterton who wrote:

> *There exists in such a case a certain institution or law; let us say, for the sake of simplicity, a fence or gate erected across a road. The more modern type of reformer goes gaily up to it and says, "I don't see the use of this; let us clear it away." To which the more intelligent type of reformer will do well to answer: "If you don't see the use of it, I certainly won't let you clear it away. Go away and think. Then, when you can come back and tell me that you do see the use of it, I may allow you to destroy it.* [Chesterton09]

---

[20] Flume, which is basically a better MapReduce.

Figure 2.2. Chesterton's fence.

The idea behind the Chesterton's fence principle is rather simple—don't remove a fence until you know why it was erected in the first place. It may have a purpose that's not obvious or visible to us, so it's important to stay intellectually humble and curious, and seek to understand before we act. Not that we should necessarily accept the status quo, but we should be aware of the trade-offs and risks involved in changing something and have a good reason for doing so.

> Don't remove a fence until you know why it was erected in the first place.

Interestingly, going back to that Google story, the commit message "You don't really need all of this stuff anymore" was mostly correct. Were those batch jobs rewritten from scratch, they wouldn't need all that excessive configuration around data center choices because the batch jobs framework already had a decent automation in place. However, these were legacy jobs written many years ago with some smart optimizations. Was it okay that there were no comments in config files explaining those optimizations? Definitely not! It's missing documentation that introduced the problem of unknown unknowns in the first place. But at the same time, was it okay to simply remove that code without trying to understand it first? The answer to that question is also no. At the very least, if after doing proper research we still had no clue, we could have tested this change on a single job, not all of them at once.

This real-life story illustrates how unknown unknowns may ultimately force us to rely on luck when making changes. But it also teaches us to always seek to understand first. Don't forget Chesterton's fence!

2.3. Complexity causes

Earlier in this chapter, we saw how Dr. Ousterhout's complexity model simplifies reasoning by making a distinction between the three complexity problems: change amplification, cognitive load, and unknown unknowns. However, when it comes to complexity causes, Dr. Ousterhout identifies only two of them: dependencies and obscurity. While I believe it is possible to track down all the complexity problems to these two, "dependencies" as a complexity cause sounds a bit vague. In this book, I propose using the following list of complexity causes:[21]

- Obscurity
- Duplication
- Unfamiliarity
- Too many dependencies
- Unstable dependencies
- Untrue interfaces
- Complex interfaces

---

[21] They're mostly orthogonal to each other; however, obscurity is certainly one of the reasons interfaces can be complex.

The rest of the chapter is structured around these seven causes of complexity: two of them have their own sections, and others are introduced in Section 2.6 "Modular design and associated complexity causes." As you may have already noticed, like Dr. Ousterhout, I don't use the term "coupling" much and instead stick to the word "dependency," which is used in a broader sense: "A dependency exists when a given piece of code cannot be understood and modified in isolation; the code relates in some way to other code, and the other code must be considered and/or modified if the given code is changed" [Ousterhout18]. Just keep in mind that what we'll call "dependency" is more commonly referred to as "coupling" nowadays.

Different complexity causes may create different complexity problems (see Figure 2.3; some of the arrows there are dotted for readability purposes). As with any other multifactor optimization, trying to eliminate one of the causes in isolation can make it worse overall. For example, in Chapter 5: "Independence Quest," we'll see how trying to minimize the number of dependencies can easily complicate interfaces and increase obscurity. So, it's important to keep all these complexity causes in mind and prioritize them based on the severity of the practical problems they create.

> Prioritize complexity causes based on the severity of the practical problems they create.

To put it positively, a healthy code is healthy because it:

- Looks *familiar*[22]
- Has little to no *obscurity*
- Has little to no *duplication*
- Has a *reasonable number* of *stable* dependencies with *true* and *simple* interfaces

[22] Yes, that's the most controversial part, see the next section for details.

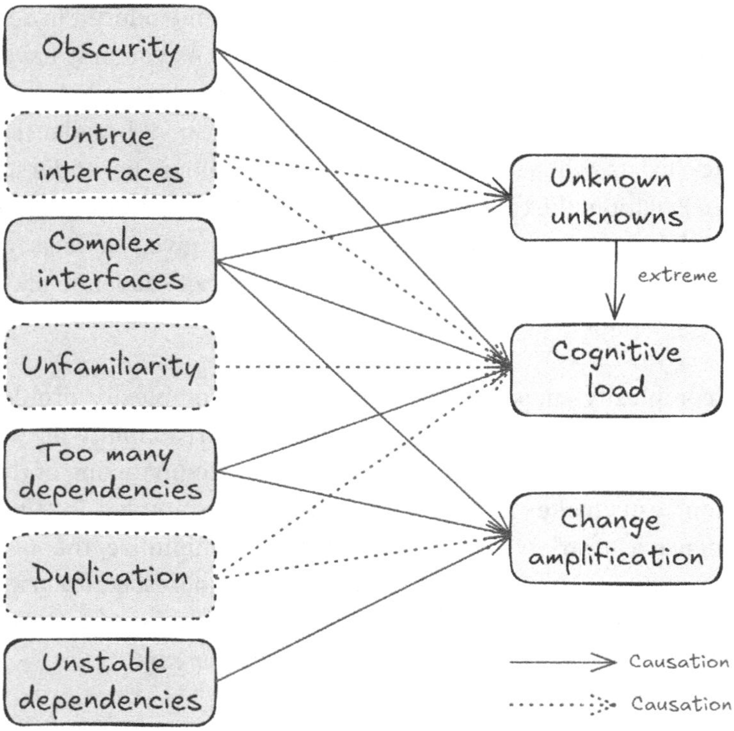

Note: Some complexity causes and their respective arrows are dotted, otherwise all the arrows become unreadable.

Figure 2.3. Complexity problems that may be created by different complexity causes.

Pause and Think

2.4. Unfamiliarity

Let's start with the most controversial complexity cause. Strictly speaking, unfamiliar syntax, patterns, or dependencies are not necessarily complex. However, they definitely contribute to the *perceived* complexity.[23] Between 2015 and 2019, Google's codebase quality perception by its engineers was in a constant decline. Yet it doesn't mean that the code health was actually degrading. In fact, the number of software engineers at Google literally doubled in that period, meaning that for many of them, Google's codebase was something new and unknown. Again, technically speaking, novelty is not complexity, but given that it's *perceived* complexity that really matters, we'll pragmatically keep unfamiliarity on our list of complexity causes.

One of the easiest ways of making your code look familiar to other developers is the use of widely known terms, syntax, patterns, and dependencies—ideally, widely known in the industry as a whole or even just within your company (see Chapter 7: "Naming: The Obscurity Champion" and Chapter 9: "Build vs. Buy").

Yet another way to make your code look more familiar is keeping it consistent. When you see `RADIUS` in C or Java code, you know that's a constant, right? Ideally, there would be a consensus in every language community around such stylistic nuances so that all the codebases would use the same code style for the same language. But at least within the same codebase, there should be nothing surprising about such stylistic nuances as line breaks, extra spaces, and curly braces. We all have our own preferences, but to be honest, it doesn't make much difference in terms of code complexity as long as you're consistent. That's why each company should have a style guide for each of the languages used and stick to it.[24]

---

[23] Those of you familiar with Rich Hickey's "Simple Made Easy" talk would probably call perceived complexity *hardness*, but for a broader audience, the term *perceived complexity* probably communicates subjectivity better.
`https://www.infoq.com/presentations/Simple-Made-Easy/`

[24] There is no need to invent your own—just adopt one of the existing style guides. For example, see `https://google.github.io/styleguide/javaguide.html`

That said, if you go beyond the code style, it's easy to take the idea of consistency too far. As soon as you attempt to regulate the quality of your codebase with a similar approach, you go down a slippery slope, as there's not much you can say in general about the right syntax, patterns, or libraries to use without having a context. Developers should be free to use their best judgment to produce the simplest code possible in each individual scenario, while following a code style and a minimal set of additional conventions widely known within a company.

> Striving for consistency shouldn't stop you from gradually adopting better practices.

Yet another problem with having strict rules beyond code style is that they hinder gradual improvements. Developers should be free to experiment and gradually adopt a better way of doing something if there is one. I'm a strong believer in the power of experiments, continuous learning, and ceaseless refactoring. If a new way of doing something is simpler, it will quickly outperform the old one in terms of familiarity. For example, you may be used to unit test names that look like:

```
linkLoginEnabledOauthAccountUsedForLoginByAnotherUserFails()
```

Even though such names do a good job at describing the tested scenario, there's definitely room for improvement. Logically separating different parts of the name with "_" reads much better:

```
linkLoginEnabledOauthAccount_usedForLoginByAnotherUser_fail()
```

There's no real problem if some test names include "_" and others do not. Both old and new naming conventions can be familiar to developers at the same time. Striving for consistency shouldn't stop you from gradually adopting better practices. Guidelines exploring different options and suggesting certain approaches are definitely good, but typically, such guidelines should avoid using verbs such as *must* and *should*, and use the more appropriate *may* and *consider* instead. That's because what a simple code looks like is highly context specific. Thus, although consistency does help with familiarity, it's easy to take it too far.

Finally, while unfamiliarity contributes to the perceived complexity, there is one more connection between these concepts: complex things rarely become *truly* familiar.

Pause and Think

A complex piece of code may live in your codebase for years, and you may even understand it at some point, but then all that sense of familiarity is quickly gone once you focus on something else. I'm sure you can relate to that, because only simple things can become truly familiar to us and benefit from our brains' cache. While unfamiliar at first, a new API has the potential to become familiar over time if it's simple. But if not, using it dozens of times is unlikely to help much in that respect. I was actually tempted to name this complexity cause "novelty" instead of "unfamiliarity." It's a shorter and stickier option, but the problem is it's misleading—something doesn't have to be novel to be unfamiliar. It can just be complex and stay unfamiliar forever.

> Something doesn't have to be novel to be unfamiliar. It can just be complex and stay unfamiliar forever.

2.5. Obscurity

The second cause of code complexity is *obscurity*. It occurs when important information isn't obvious [Ousterhout18]. In this book, most causes of complexity have their own chapter. For instance, Chapter 3 focuses on *untrue interfaces*, Chapter 4 on *complex interfaces*, and Chapter 5 on *too many dependencies* and *duplication*. Obscurity, however, stands out and is discussed in *most* of the following chapters. Such extensive coverage is not only because of its importance but also because obscurity blends so naturally with other causes of complexity. Thus, the landscape of different forms of obscurity in code is incredibly diverse.

Naming. For example, *naming* can be obscure, and for many different reasons. The noblest one of them is keeping names short (see Chapter 7: "Naming: The Obscurity Champion").

Coupling. Even the very *fact of having a dependency* may be obscure. How many times have you found yourself fixing some seemingly unrelated code that was broken by your changes? Pragmatically speaking, coupling by itself isn't that bad—it is *implicit* (or, in other words, obscure) coupling that is a real hazard. For example, consider the following comment in a function body:

```
// Skip the access checks here as the caller makes all the
// necessary access checks on their side.
foo();
```

At first glance, this code snippet looks pretty innocent and even good as it explains its assumptions, right? But if that's the only comment left, there is a huge problem on the other side—working on the calling code you wouldn't know of these assumptions![25] We'll discuss making coupling explicit in Chapter 5: "Independence Quest."

Documentation. After naming and implicit coupling, *lack of or misleading documentation* is probably the next most common source of obscurity. It may be some high-level standalone documentation (aka wiki) or a contract that is missing, incomplete, or plainly wrong. We'll talk about contracts in Chapter 3: "Design by Contract," and even though there is no chapter on wiki,

[25] Yes, some of us would guess, and those who worked with this code before might still remember.

there is Chapter 8: "Trade-off Analysis and Design Docs." Although strictly speaking, design docs are not documentation, they also draw a bigger picture, addressing the problem of high-level obscurity.

Comments. Furthermore, some people believe that good code doesn't need any *inline comments*. That is totally true if you want to know what the code is doing or where it's being used within the codebase—duplicating such knowledge in comments is indeed redundant, violates the DRY principle (see Section 5.1), and typically becomes confusing in the long run as developers tend to forget to update such comments. Yet the code by itself doesn't explain *why* it's doing what it is doing,[26] so comments answering the *why* question are orthogonal and complementary to the code itself, and in some cases, drastically improve readability.

Pause and Think

Code that communicates your doubts

The main idea behind Donald Knuth's literate programming methodology is to communicate the human thought process—to put humans first instead of focusing merely on executable computer code. That's what the comments answering the *why* question are about—communicating the thought process. Yet another missing component here is our doubts. We're not always sure that the code we've written describes our intentions in the best way possible, right? But without any comments, it reads as if the author was confident in what they wrote. So I find leaving comments about my doubts (and alternative approaches I considered) extremely helpful. They're not necessarily answering the *why* question but are definitely communicating my thought process, very much in the spirit of literate programming.

Logs, traces, and metrics. When you're debugging an issue and reading the logs, it's not uncommon to feel that "important information isn't obvious" (which is the definition of obscurity). There may be an important *log message*

[26] That's especially true for performance optimizations.

missing, or the log message may be there but missing some important details about the internal state of the program. Similarly, lack of traces or metrics may also increase obscurity around what was actually happening under the hood.

Mutability. Often it's debugging that reminds us of a typically overlooked code complexity dimension—a temporal one. When code actually executes, there may be billions of possible execution states and scenarios, primarily because of the time dimension and *mutable data,* which is another super common source of obscurity. Which instruction has changed that value? And when? Today, there's kind of a consensus in the industry that the global state is evil. As *The Pragmatic Programmer* puts it, "each piece of global data acts as if every method in your application suddenly gained an additional parameter" [HT19]. Having that consensus is great, but I believe there is room for improvement here—it's avoiding any mutability as much as possible (see Chapter 6: "Semi-functional Programming").

Other. It's a similar story with *cache invalidation* and *eventually consistent data replication*—is the data you're working with still valid or already stale? More specific technologies may also have their own sources of obscurity, such as data obscurity in schemaless NoSQL databases, obscurity around causation in publisher–subscriber pattern in event-driven design, and so forth.

Some of this obscurity may be a necessary trade-off, but my goal here is to bring your attention to the fact that all of the above is obscurity, which causes the problem of unknown unknowns, the worst code complexity problem.

2.6. Modular design and associated complexity causes

The next complexity cause is *duplication.* Even though special in some ways, essentially it's an extreme form of a dependency. A piece of code depends on its duplicate, and its duplicate in turn depends on the original. What makes duplicates worse than a typical dependency is that there isn't always an ideally symmetric bidirectional dependency between them. We'll talk about it in detail in Section 5.1 "Don't Repeat Yourself."

Unlike the first three complexity causes (unfamiliarity, obscurity, and duplication), the rest of them only exist in the context of modular design. So let's introduce a couple of definitions first. For the purposes of this book, any unit of code that has an interface and implementation is called a *module*. A module's interface gives a promise, and its implementation carries it out. Per this definition, functions and classes are obviously modules. But it's not only them—microservices, libraries, and frameworks also have an interface and implementation parts. So per this definition, modern software systems are modular at every level (see Figure 2.4).

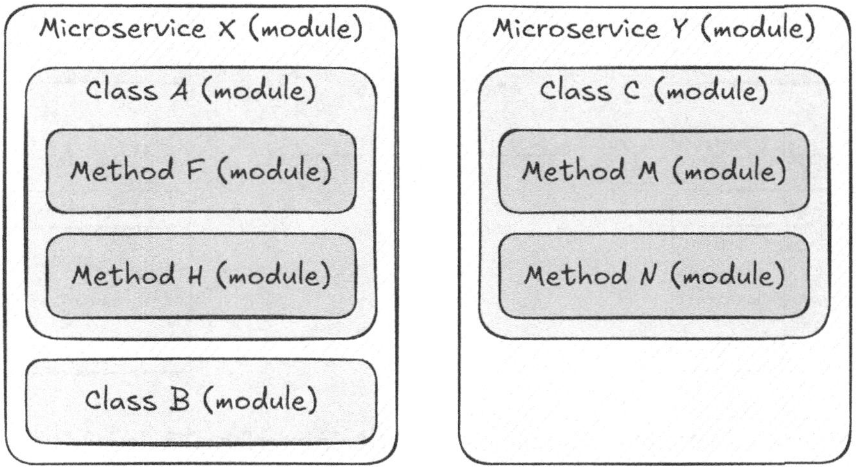

Figure 2.4. Sample backend decomposition into modules.

Modular design is as old as programming itself. Procedures (called subroutines in the past) were one of the first inventions in computer science, paving the road for modular design as a whole. The key idea behind it is rather simple: to use a module, you don't need to understand how it works under the hood, you just need to understand its interface. Whenever you read `map.put(key, value)` or `dictionary.put(key, value)`, you don't need to remember or even understand the details of hash table implementation, you only need to remember this simple interface.

Given that modules are so ubiquitous in software engineering, it's no surprise that peculiarities of a system's decomposition into modules have a huge impact on its overall complexity. Bad decomposition into modules can be truly disastrous. Looking at Figures 2.3 and 2.5, it's easy to see that all the possible causes of complexity are at play here.

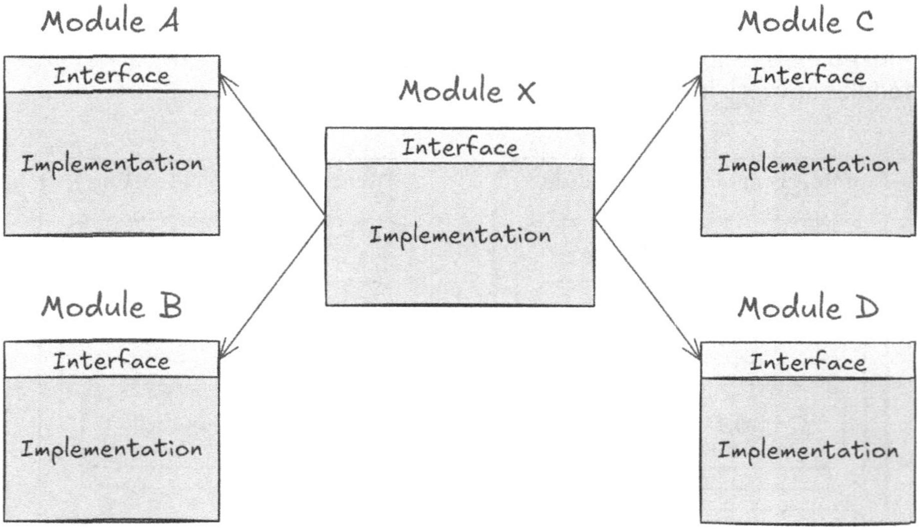

Figure 2.5. Module X and the modules (their interfaces) it depends on.

Looking at Figure 2.5, there is one important observation to make: any dependency is a dependency on the module's interface, not a dependency on the module itself.[27]

Pause and Think

[27] Once an interface becomes popular enough, in practice, as an owner of the module, you can't rely on this statement anymore (see Hyrum's Law), but as a client, you should always be thinking about it this way. `https://www.hyrumslaw.com/`

Now that the modular design concept is well-defined, let's take a look at the four remaining complexity causes.[28] We'll take a look briefly, leaving the detailed discussion to other chapters.

Too many dependencies. It's a scientific fact that we have a single-digit limit on how many items our brains can operate with at the same time.[29] It is called "working memory" in psychology, and in the first approximation, you can think of it as a small number of CPU registers. So, if our goal is code readability (thinking about our fellow humans), we can't ignore this fact—we must make sure there are not too many dependencies at every level. Chapter 5: "Independence Quest" is built around exploring the "fewer dependencies" idea. Also, Chapter 6: "Semi-functional Programming" demonstrates some extremely effective ways of having fewer dependencies, while staying within the OOP[30] paradigm.

Unstable dependencies. If a library you use turns out to be not backward compatible with your usage, a web service you depend on suddenly becomes unavailable, or there is a security hole yet to be discovered in one of your dependencies, all that instability can lead to change amplification problems (see Chapter 9: "Build vs. Buy"). Yet unstable dependencies are not necessarily external ones—it may be your company's internal service or your own code that is unstable. In fact, if you're in the active development phase, your changes will rarely be backward compatible, right? However, it is possible to make your interfaces more stable by:

- Finding the right abstractions (see Section 5.3)
- Avoiding dysfunctional layers (see Section 5.5)
- Making interfaces simple, as simple interfaces are much less susceptible to change than their sophisticated counterparts (see Chapter 4: "Interface Simplicity").

---

[28] That doesn't mean that the first three causes are never dependency related. Of course, dependencies can be obscure or unfamiliar.
[29] Nelson Cowan, "The Magical Mystery Four: How is Working Memory Capacity Limited, and Why?" `https://www.ncbi.nlm.nih.gov/pmc/articles/PMC2864034/`
[30] Object-Oriented Programming

Untrue interfaces. The name of the next complexity cause, "untrue interfaces," is easy to misinterpret. Such an emphasis on the importance of having correct interfaces doesn't mean we don't care about the correctness of all the other code! We definitely want all the code to be correct, but given that interfaces are the backbone of modular design, it's *especially* important for them to be complete[31] and trustworthy. That's the primary topic of Chapter 3: "Design by Contract."

Complex interfaces. If you want your development efforts to scale, you need the interfaces used to be not only true, but also simple. Similar to interface correctness, of course we want *all* the code to be as simple as possible, but for the very same reasons as above, it's *especially* important for interfaces. For most projects, it would be unrealistic to aim at having a truly simple codebase as a whole; however, simple interfaces are not only realistic, but also an indispensable part of the story of a healthy codebase. There is always some unavoidable complexity in code, and hiding it under a simple interface is the best way to deal with it. I believe the following to be one of the most fundamental facts in programming—"isolating complexity in a place where it will never be seen is almost as good as eliminating the complexity entirely" [Ousterhout18]. In other words, not all complexity is born equal. Having a complex interface is typically much worse than having a complex implementation (see Chapter 4: "Interface Simplicity").

Interestingly, at first glance, having complex interfaces as a complexity cause may look like a circular reference. Want to avoid complexity problems? Well, avoid complexity. But in fact, interface complexity is much easier to define than code complexity in general. Even if you cannot articulate it, you definitely have an intuition for what constitutes a simple interface.

To avoid all six complexity causes, in good modular design we strive to have:

---

[31] Isn't being incomplete also a form of untruth? This philosophical question dates back to Satan tempting Eve in the Garden of Eden.

- *Fewer non-trivial dependencies* and *duplication*
- Dependencies that are *explicit, familiar*, and *stable*
- Interfaces that are *true* and *simple*[32]

Finally, while it's rather obvious that all these ideas have a *potential* to simplify any particular system, that doesn't mean they actually will. Some of them are not as harmless as they may seem. As always, the devil is in the details. Complexity-wise, modular design is a double-edged sword, and in the rest of the book, we'll be learning how to handle it effectively and safely.

Summary

Code complexity is not a single problem. In fact, there are three problems: change amplification, cognitive load, and unknown unknowns. Change amplification is the least critical one as it typically doesn't lead to exhaustion or bugs—it just slows you down. Unknown unknowns is the worst one as it makes you rely on luck when making changes.

Healthy code is code that looks familiar, has little to no obscurity and duplication, and has a reasonable number of stable dependencies with true and simple interfaces. Same as complexity problems themselves, not all the causes of complexity are equally bad. They should be prioritized based on the severity of the complexity problems they create.

Finally, somewhat counterintuitively, a healthy codebase may have quite a lot of complex code. That's because "isolating complexity in a place where it will never be seen is almost as good as eliminating that complexity entirely" [Ousterhout18].

---

[32] Note that some of these ideas work in both directions and even have a boomerang effect. For example, to minimize a module's complexity, we want it to depend on other modules with simple interfaces, but at the same time, we want its own interface to be simple as well so that the other modules may be simplified.

The code complexity model introduced in this chapter will be assumed and extensively used in the chapters that follow. Yet note that the order of the following chapters generally doesn't matter, so feel free to read them in an order that's most meaningful to you.

Chapter 3

DESIGN BY CONTRACT

Modular design and encapsulation are the bread and butter of software engineering. However, like any powerful and flexible technique, they can be harmful. An extra module can increase code coupling instead of decreasing it, or the interface exposed may be too complex, incomplete, or even plainly misleading. In the next three chapters, we explore the pitfalls of modular design, starting with its most fundamental problem—building trust in modules' interfaces. The first part of this chapter introduces the concept of contracts, and the second one provides an overview of relevant documentation practices and contract enforcement mechanisms. To build trust in modules' interfaces, their contracts must be properly documented[33] and enforced.

The ultimate goal of this chapter is to make it crystal clear that Design by Contract is not one of the myriad optional techniques that developers may or may not adopt, depending on their school of thought, but rather an essential ingredient of any nontrivial healthy codebase.

3.1. The interface trust dilemma

Trust is existentially important for any successful relationship, whether it's among family members, friends, business partners, or your team. It is true that without trust relationships are still possible; however, they become complicated quickly. Somewhat surprisingly, this is also very true for

[33] Which sometimes means no documentation at all; see Section 3.3.

relationships among software modules. Code that doesn't trust the interfaces of other modules can rarely boast good readability. It often looks something like this:[34]

```
void foo(User son) {
  String parentName = null;
  if (son != null) {
    try {
      var parent = son.getParent();
      if (parent != null) {
        parentName = parent.getName();
      }
    } catch (Exception e) {
      // TODO: try to get parent's name from DB.
    }
  }

  if (parentName == null || parentName.isEmpty())
  // ...
}
```

Instead of such spaghetti code, it's supposed to be simply

```
void foo(User son) {
  var parent = checkNotNull(son.getParent());
  // just use parent.getName() below
}
```

However, such ridiculous safety nets are not necessarily paranoid. Unfortunately, for many of us, that's what our past experience tells us to do—have a plan B! So, there is a tension between the two forces here. On the one hand, modules should depend solely on the other modules' interfaces (see Figure 2.5). Understanding a module's interface should always be enough to use it—that's the key premise of modular design. On the other hand, it's not

---

[34] Yes, it may be a little bit exaggerated.

that uncommon for an interface to be unclear about the module's intentions, or even worse, to communicate the intentions clearly but de facto do something else. Thus, the temptation to add such safety nets is absolutely understandable, yet counterproductive (see Figure 1.1). Even if such spaghetti code works, you never know why, and "if you're not sure why it works, you won't know why it fails" [HT19].

> Understanding a module's interface should always be enough to use it.

3.2. The power of enforced contracts

Whether to trust interfaces is a terrible dilemma to have, and there's only one good solution to it—enforced contracts. To make your modules' interfaces trustworthy,[35] they should always be *well-documented* and *internally enforced*. That may sound a bit idealistic, but first, there is no other way to have robust simple code, and second, as we'll see further down in this chapter, following these two rules in practice is not that hard.

In essence, every module's interface is a contract. Similar to business contracts, it has benefits and obligations: those of the module itself and of its callers. For example, consider a `bankAccount.deposit(amount)` method. The contract here is that if a bank account is open, and the amount being deposited is positive, as a result of a method call that amount would be deposited into the account. From the callers' perspective, preconditions such as "bank account must be open" are their obligations. But for the module itself, these constraints are clearly benefits, as they define the implementation boundaries—now there is no need to handle negative amounts as credit operations or initiate (re)opening an account if it's currently closed (that would be scary, right?). In return, if the caller meets its obligations, the module promises to do the

[35] At the very least in your codebase.

transaction, which is the module's obligation and the caller's benefit. Explicit or implicit, such obligations and benefits are the very crux of any interface. If not documented, they are definitely assumed.

However, having a contract by itself doesn't make a huge difference. To be truly beneficial, contracts have to be clear, and they have to be enforced. That's what Design by Contract technique is about—building trust in modules' contracts by ensuring that they are always *well-documented* and *internally enforced.*

As you can rightfully expect from the only solution to such a foundational problem as the interface trust dilemma, Design by Contract technique is not a novel idea. Although the term itself was coined by Bertrand Meyer in the late 1980s [Meyer88], it all goes back to Hoare triples, proposed by Tony Hoare in 1969 [Hoare69]. It's a good old idea really. But even more importantly, it's also a rather obvious one:

- *Documenting* what a module expects (preconditions) and what it does (postconditions and invariants preserved) is at the very heart of modular design.
- *Enforcing the contract internally* is very much in the spirit of encapsulation—a module should encapsulate not only its contract's implementation but its enforcement as well.

And that's all it takes to make interfaces trustworthy! If you practice these two rules consistently within your codebase, you'll notice that over time, it builds trust in your interfaces among developers, serving as a solid foundation for keeping your codebase healthy.

3.3. Document contracts

There are three types of problems with documentation to avoid:

- Missing documentation
- Misleading documentation
- Redundant documentation

Interestingly, most interfaces don't need much documentation. To see why, let's document the contract of the deposit method discussed above (ignoring error handling for the sake of simplicity):

```
interface BankAccount {

  /**
   * Note that the account must be open.
   *
   * @param[36] amount Should be positive.
   */
  void deposit(int amount);
```

Note that this documentation has the callers' obligations but not their benefits. That's because the method's name already communicates those—as a result of the call, the specified `amount` would be deposited into the account. But we could have also communicated the two preconditions in exactly the same way: a more specific `positiveAmount` parameter name would make the "Amount should be positive" comment redundant, and using an `OpenBankAccount` subclass would make the second comment redundant, so that `openBankAccount.deposit(positiveAmount)` doesn't need any of the formal documentation anymore to communicate its contract!

---

36 The `@param` tag in Java is used to document parameters.

> The interface is well-documented if documentation has all that is missing in code that may be necessary to use the module, and nothing else.

It may be a synthetic example, but it illustrates perfectly the point that the "well-documented contract" requirement is not about excessive documentation. The interface is well-documented if documentation has all that is missing in code that may be necessary to use the module, and nothing else. In other words, documentation should not only be complete, but also concise. There is no value in documenting facts that are already present in code, and there are extra risks associated with doing so, because such documentation tends to go out-of-sync with reality over time and eventually becomes utterly confusing.

Don't be pedantic with documentation

One of the first internal Google practices I learned after joining was that I don't have to be pedantic and document every function parameter. That had such a liberating effect on me because the Javadoc documentation was saying exactly the opposite: "`@param` tag is 'required' (by convention) for every parameter, even when the description is obvious."[37] Whenever you violate that convention, your IDE may even show a warning by default. But if you think about it, it's absolutely okay and even desirable to skip all the redundant `@param` tags and have only those that actually clarify the contract:

```
/**
 * @param state Repo state will be updated as a result of the call.
 */
void syncGitHubRepo(RepoState state, GitHubUser myGitHubUser);
```

37 https://www.oracle.com/au/technical-resources/articles/java/javadoc-tool.html

Some could argue that in this particular context, it was already obvious that `amount` should be positive and that `bankAccount` should be open, which would imply that what we started with *has already been communicating* its contract properly as there were no nonobvious details to document:

```
interface BankAccount {

  void deposit(int amount);

  // ...
}
```

Yet what is obvious and what is not depends on whom we ask. If you're 100% certain that these nuances will be obvious to everyone who will ever read this code, then sure, there is no need to make them explicit. Neither in code nor in documentation. But if in doubt, don't leave it out.

Contracts must be clear. Without a clear contract, decomposition into modules is not much better than

```
void foo(params) {
  fooPart1(params);
  fooPart2(params);
}
```

This may achieve the goal of having functions no longer than 50 lines of code, but obviously, it doesn't accomplish anything truly valuable. Don't tolerate vague promises or constraints in interfaces. Whenever you spot them and can fix them, please do so! But also consider going through the module's current usages within your codebase to check that its contract hasn't been already misinterpreted (by somebody else or even by yourself).

Finally, same as with code complexity in general, an unclear, wordy, or incomplete contract is usually a product of many small incremental changes. It starts as a reasonably understandable interface, but then time and inertia convert it into what it is today. So whenever you make a code change, ask yourself whether the contracts of the touched modules are still in good shape. Keeping contracts clear, concise, and complete is a continuous effort and a shared responsibility.

3.4. Enforce contracts with tests

Having a clear contract is a good start, but usually, contracts are pretty useless unless properly enforced. Just take a step back and look at the world today—businesses flourish in countries with solid law enforcement practices, right? A country may have perfect legislation, but a broken law enforcement system would make it too unpredictable for most businesses to operate there. To have a real value in the real world, laws and contracts must be enforced. Interestingly, that's very true for software modules' contracts as well.

One of my ex-managers seemed to be obsessed with asking during incident reviews whether any test that would have prevented the incident from happening was missing. But it's a very good question to ask, given that automated tests are the primary mechanism of enforcing software correctness. And that is true for software modules at every level, regardless of whether we're talking about a UI component, microservice, standalone utility, or just an individual function. Want to make sure a software module does what its interface promises to do? Test it![38]

---

[38] Obviously, that's the responsibility of the module's owner to have all the necessary tests in place. As mentioned above, enforcing the contract internally is very much in the spirit of encapsulation—a module should encapsulate not only its contract's implementation but its enforcement as well.

Master testing

Several years ago, the book *Software Engineering at Google* [WMW20] was published, sharing the lessons learned from programming over the two decades of building Google. As you would rightfully expect, there's a lot of wisdom there, but pragmatically speaking, I wouldn't recommend reading it cover to cover unless you're in a leadership position, as it's more about building a software engineering organization than about anything else. However, there are some chapters for a broader audience. Specifically, there are brilliant "Testing Overview" and "Unit Testing" chapters. This section draws heavily from them, but I definitely recommend reading those chapters in full. The book is available online for free at `https://abseil.io/resources/swe-book`. Unlike writing good code in general, writing good unit tests is not that hard. It's a shame that so many programmers (even senior ones) haven't mastered it yet.

A good test is clear, complete, and concise. Like contracts, it has all the details necessary to understand the scenario under the test and nothing else. Just reading the code of the test top to bottom should be enough to understand why it should pass, without any extra lookups. Given that we don't write tests for tests, they must be *obviously* correct. Here's an example of a test that is neither complete nor concise:[39]

```
@Test
public void twoPlusThreeMustBeFive() {
  Calculator calculator =
      new Calculator(
          new RoundingStrategy(),
          "Unused",
          ENABLE_COSINE_FEATURE,
          0.01,
          calculusEngine,
          false);

  var result = calculator.calculate(newTestCalculation());
```

[39] Borrowed from the "Unit Testing" chapter of *Software Engineering at Google* [WMW20].

```
  assertThat(result).isEqualTo(5);
  // Where did this number come from???
}
```

This test doesn't have all the information needed to understand why it should pass, and it also has a lot of unnecessary details. Here is a complete and concise version of the same test:

```
@Test
public void twoPlusThreeMustBeFive() {
  Calculator calculator = newCalculator();
  var result =
      calculator.calculate(
          newCalculation(2, Operation.PLUS, 3));
  assertThat(result).isEqualTo(5);
}
```

All the unrelated details are hidden. Everything that matters is explicit. Again, good tests are complete and concise.

Tests as another reason for simple contracts

It's almost impossible to write a clear test if the contract under the test is not clear itself.[40] Interestingly, if lack of documentation is the only problem with the contract, then a good test suite would even mitigate that problem to some extent—tests serving as a kind of documentation. However, if the module's contract is de facto complex, then tests are almost guaranteed to be confusing, too. We'll talk more about it in Chapter 4: "Interface Simplicity."

[40] Yes, in some rare cases, that's unavoidable (e.g., the contract of a JPEG encoder is by definition super complex). In cases when you can't objectively have intellectual control over the contract (or over the inputs), you would need some advanced forms of testing such as fuzzing or property-based testing. See a brilliant "Design is Testability" tech talk by Titus Winters on this topic. `https://www.youtube.com/watch?v=eBOn1e113Ak`

Finally, while it's extremely important for tests to be clear, it's even more important for them to actually verify the correctness of the code under the test. That's exactly why they exist in the first place. Surprisingly many tests don't do so—specifically those tests that are abusing mocking frameworks. Here is a quote from *Software Engineering at Google*:

> *When mocking frameworks first came into use at Google, they seemed like a hammer fit for every nail—they made it very easy to write highly focused tests against isolated pieces of code without having to worry about how to construct the dependencies of that code. It wasn't until several years and countless tests later that we began to realize the cost of such tests: though these tests were easy to write, we suffered greatly given that they required constant effort to maintain while rarely finding bugs.* [WMW20]

The main problem with tests that use mocking frameworks is that they don't truly enforce the contracts. Today, internal Google guidelines strongly recommend making tests as realistic as possible, that is, to use real objects in tests unless they're slow or nondeterministic [WMW20, "Unit Testing"]. If the real objects are too slow or nondeterministic, use a fake[41] instead, and only in the absence of a realistic fake should you use a mocking framework as a last resort.[42,43]

Pause and Think

---

[41] Interface owner is the one who is responsible for creating realistic fakes for testing [WMW20].
[42] And use a stub (commonly inaccurately referred to as mock) for testing dependency failure scenarios [WMW20, "Test Doubles"].
[43] Despite interaction tests being brittle, sometimes testing state only is not enough. For example, if the correctness of an API depends on things like "X is never logged," "`UpdateUI()` is called exactly one time," or "we issue at most three outbound RPCs," those types of things are only handled with a mock. `https://abseil.io/resources/swe-book/html/ch12.html#test_statecomma_not_interactions`

Lesson learned the hard way

Obviously, having clear internal guidelines against mocking frameworks abuse didn't automatically solve that problem at Google. Specifically, in our team, we've been abusing mocking frameworks for years, being unaware of those guidelines. Later, I've heard from my teammates that they also had a gut feeling something was wrong with our tests. They were unnecessarily brittle (that is, failing without a good reason, just because of some unrelated changes) and could have done a better job at catching bugs. But I guess it was pretty hard for us to challenge the status quo. So, it was great to discover those guidelines on our intranet, although it required a certain amount of luck (*Software Engineering at Google* [WMW20] wasn't published yet, and honestly, I wouldn't be surprised if some teams at Google still haven't read the book and are unaware of these guidelines). Knowledge sharing, even within a company, is hard.

In light of these recommendations, it becomes almost impossible to draw a line between a unit test and an integration test, but that's probably okay. Don't be misled by the word "unit" in the phrase "unit tests"—don't mock out everything else apart from the class under the test! What is *really* important is that a realistic test is not brittle and actually enforces the contract of the module under the test.[44]

> Don't be misled by the word "unit" in the phrase "unit tests"—don't mock out everything else apart from the class under the test!

[44] A realistic test also helps enforce the contracts of the modules used by its implementation. Yes, it is a primary responsibility of those modules themselves to verify their contracts, but some assistance from the outside at no extra cost is also very much appreciated.

3.5. Enforce contracts with assertions

While tests are the primary means of enforcing software correctness, there is also another powerful but often overlooked tool—assertions, also known as runtime verification checks.[45] More specifically, in the context of contract enforcement, we're interested in preconditions, which are a special type of assertions that validates that the caller obeyed the requirements of the module's contract. Here is how precondition checks would look for the `bankAccount.deposit(positiveAmount)` method we discussed before:

```
/**
 * @throws IllegalArgumentException if positiveAmount is not
 *       positive or if this account is not open.
 */
void deposit(int positiveAmount) {
  checkArgument(
      positiveAmount > 0,
      "Amount must be positive but found [%s].",
      positiveAmount);
  checkArgument46(
      this.state == State.OPEN,
      "Bank account [%s] must be open but found [%s].",
      this, this.state);
  // ...
}
```

If the caller hasn't met their obligations, a precondition check just crashes the program. While failures in runtime can be highly unpleasant,[47] the alternative

[45] Don't confuse them with `assert(...)` statements in Java, as that implementation of assertions has several serious downsides. So to avoid confusion you would often refer to assertions as "runtime verification checks." It's not rare to have some flexibility in how to implement them in different languages. For example, in Go, that's not necessarily panic; they may be implemented as structured errors as well.

[46] Java experts, yes, it should be `Preconditions#checkState` instead.

[47] And there are certain rare cases (types of applications) in which you shouldn't fail in runtime.

is even worse—if you proceed with a negative amount to deposit into the bank account, prepare for surprising results later on. As *The Pragmatic Programmer* puts it:

> *Crashing is often the best thing you can do. The alternative may be to continue, writing corrupted data to some vital database or commanding the washing machine into its twentieth consecutive spin cycle. [...] When your code discovers that something that was supposed to be impossible just happened, your program is no longer viable. Anything it does from this point forward becomes suspect, so terminate it as soon as possible. A dead program normally does a lot less damage than a crippled one.* [HT19]

Use assertions only for checking things that should really never happen

Assertions is a tool for catching bugs, such as callers violating the requirements of the module's contract, for example. Asserting something that must be true in 99.99% of cases only won't end up well. And it's not only about user input validation. As another example, I would also totally abstain from assertions in a job that analyzes a database dump—just log and count the number of anomalies instead. Unless some invariant is enforced by the database schema, you can almost guarantee it's not always preserved. You don't want the entire job to fail because of a data anomaly in one of the rows.

Probably the most popular reason for avoiding assertions is a common belief that well-tested programs shouldn't have bugs, and if there *are* bugs, it means there is a problem with test coverage. *The Pragmatic Programmer* responds to that brilliantly:

> *First, for any complex program you are unlikely to test even a minuscule percentage of the permutations your code will be put through. Second, the optimists are forgetting that your program runs in a dangerous world. During testing, rats probably won't gnaw through a communications cable, someone playing a game won't exhaust memory, and log files won't fill the storage partition.* [HT19]

Tests are your first line of defense, and assertions are the second.

> The caller's responsibilities must always be enforced by precondition checks.

Finally, note that problems discovered by assertions are not necessarily discovered in production. There are high chances of spotting a problem earlier in staging environment or in canary during the production release process.

3.6. Mitigate assertion failures

Any contract violation is a bug. The caller's responsibilities must always be enforced by precondition checks. But at the same time, there are several valuable techniques to make your production code more resilient to precondition failures:

Suppress errors from optional operations (and maybe retry them later).[48] Let's say you have a main operation followed by an optional operation. If the second one fails, you probably don't want to rollback the first one and spoil the UX[49] by showing the user an error. So pragmatically speaking, just wrap the optional operation in a try-catch, log its failure, and ignore it. In some cases, it may be fixed later by a daily sync job, or you may need to explicitly add it to some task queue to process it later. Or maybe it's a *truly* optional operation that doesn't even need to be retried.

Relax a constraint. Sometimes, contract requirements could be made less strict than they currently are. For example, consider the

[48] That's typically done by wrapping an optional operation in a try-catch, but not necessarily. Errors are suppressed differently in languages with explicit error handling, such as Rust or Go, and even in asynchronous Java code.
[49] User experience

`bankAccount.deposit(positiveAmount)` method. Why not allow a zero `amount` there and do nothing if zero is passed in? It may be an option.

Get rid of a constraint. Sometimes it's possible to get rid of a contract constraint altogether without introducing an additional implementation burden. Consider the `list.delete(element)` method. Instead of throwing an exception if the element is not present in the list it may simply do nothing (NOP). By slightly changing the method's contract, we're not only simplifying it but also making the production code more resilient.

Finally, apart from preconditions, you may choose to verify some of the postconditions and invariants in runtime as well. While precondition checks must be encapsulated within a module, postcondition and invariant checks may sometimes be done externally:

```
var sortedItems = sortItems(items);
verify(areSorted(items));
```

However, such external verification should be more of an exception. Even if you cannot modify the module in question to encapsulate its contract enforcement, you can still use a decorator pattern to do so. It is a primary responsibility of a module itself to verify its own contract.

Summary

Document and enforce contracts internally—that's all it takes to make interfaces trustworthy. If you practice these two rules consistently within your codebase, you'll notice that over time, it builds trust in your interfaces among developers and, as a result, dramatically improves the health of your code. Trustworthiness of interfaces is the first cornerstone of a healthy codebase.

Chapter 4

INTERFACE SIMPLICITY

The previous chapter demonstrated how important it is for interfaces to be trustworthy. Yet, in addition to being true, good interfaces must also be simple. That's their second property that is so essential that it deserves a separate chapter as well. Of course, interfaces are not the only code that needs to be true and simple, but as we'll see in Section 4.1, this is *especially* important for them.

Luckily, it's also much easier to have simple interfaces than to have simple code in general. Usually, saying "Our codebase is simple" is a bit of an overstatement, but having truly simple interfaces is absolutely realistic. So given the utmost importance and feasibility of interface simplicity, investing in it is a wise choice. You won't regret the time and effort spent on simplifying interfaces.

If you find this chapter to be too complicated, it is recommended to skip ahead to chapter 7. The following chapters are mostly orthogonal to each other and greatly differ in complexity.

4.1. Interface first

In the preface to the legendary *Effective Java,* Joshua Bloch mentions that because of his background in designing reusable components, he naturally thinks of interfaces first. But, interestingly, then he also suggests that regardless of whether you're consciously writing a reusable component or not, it always helps prioritizing interfaces (APIs):

> *I naturally think in terms of exported APIs, and I encourage you to do likewise. Even if you aren't developing reusable components, thinking in these terms tends to improve the quality of the software you write. Furthermore, it's not uncommon to write a reusable component without knowing it: you write something useful, share it with your buddy across the hall, and before long you have half a dozen users. At this point, you no longer have the flexibility to change the API at will and are thankful for all the effort that you put into designing the API when you first wrote the software.* [Bloch08]

If you'd ever want to estimate the complexity of a codebase, it's unlikely that cyclomatic complexity or any other formal approach would help you much. But there's a formula that might be very useful:

$$C = \sum_{p} c_p t_p, \qquad (1)$$

where C stands for complexity, p is part of the system, and t_p is the time developers spend working on that part [Ousterhout18]. Basically, this formula says that complexity of a codebase is a weighted average of complexities of its parts, where weights are the time developers spend working with those parts.[50]

Pause and Think

This is a very pragmatic formula with an interesting and almost obvious corollary: there is not much value in simplifying code that developers won't work with much! As Ousterhout puts it, "Isolating complexity in a place where it will never be seen is almost as good as eliminating the complexity entirely" [Ousterhout18].

50 Unless we're talking about potential security holes, of course.

Now, let's get back to interfaces, a part of the code that is particularly hard to isolate. An interface that is used X times across the codebase appears X times in the formula above, contributing X times to the overall complexity of the codebase. Yet the impact of interfaces on codebase complexity varies greatly, and it's not just X that matters. Basically, there are two factors:

- How often would developers bump into that particular interface, or in other words, how frequently would they need to read and understand its usages?
- How much control would one have over those usages, just in case they want to change the interface later?

Given how much things can change over time, it's often hard to guess the answers to these questions beforehand. A private API may evolve into a public one, and some parts of the codebase will definitely take an unproportional part of developers' time. But, as a rule of thumb, the importance of being true and simple increases alongside the interface visibility:

```
private method
  < public method
    < public method in a public class
      < company-wide API
        < public API[51]
```

Reevaluate interface complexity when modifying its visibility

It's worth reevaluating interface simplicity whenever you upgrade its visibility. Remember that that `private` keyword in front might have been the reason why the author didn't care about its complexity much.

Thus, the first two arguments for the interface-first mindset are their ubiquity and the fact that it can be hard or even impossible to change them later. But there's more. When thinking about a complex interface, it's probably cognitive

[51] Public API is what Joshua Block calls "exported API" in the opening quote.

load that first comes to mind (see Figure 2.3), yet in fact, complex interface is a unique antipattern that can *cause all three complexity problems at once.* To put it positively, making interfaces simple addresses all three of them (see Figure 4.1):

1. A simple interface reduces *cognitive load* by improving readability of the module itself and readability of all of its usages. Moreover, a simple interface also becomes familiar much faster, so the next time you bump into it, the cognitive load is even smaller.
2. A simple interface makes the module more findable and makes it easier to start using it. Thus addressing the two main causes of knowledge duplication, which in turn reduces the risk of *unknown unknowns* and *change amplification* problems (see Figure 5.3).
3. A simple interface is usually more stable,[52] further reducing the risk of *change amplifications.*

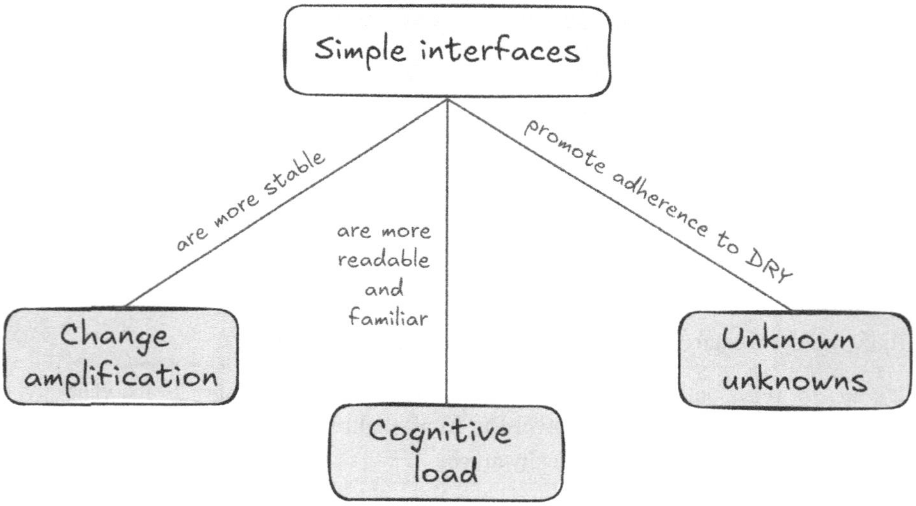

Figure 4.1. Simple interfaces address all three complexity problems.

---

[52] Just as a fact, a function with five parameters often gets the sixth one, or one of the existing parameters changes.

In summary, there are five reasons for keeping your interfaces simple:

- Their ubiquity
- Difficulty of later changes
- Fewer change amplifications
- Lower cognitive load
- Less unknown unknowns

This makes interface simplicity a second cornerstone of the healthy codebase. Together with their trustworthiness, these two cornerstones make the interface-first mindset an indispensable part of productive software engineering.

4.2. Deep and shallow interfaces

It's mostly obvious what constitutes a simple interface. For example, for functions, it's a clear name, a small number of well-named parameters, and a concise contract. It's not rocket science, right? Yet, there is one caveat. Not every interface that is easy to read is also easy to use. As an extreme example, most modern (RISC) assembler instructions are rather simple, but that doesn't make programming in assembler an easy task, even if you're working on some low-level problems such as device drivers or firmware. To write in assembler, you need to learn not only the instructions themselves, but tons of their usage patterns as well.

Pause and Think

This observation leads us to what I believe is one of the most important concepts in modular design—the concept of interface depth. Let's define it as a simple proportion:

```
InterfaceDepth = Functionality / InterfaceComplexity          (2)
```

Or, alternatively, using the contract terms from the previous chapter:

```
InterfaceDepth = ContractDepth =
CallerBenefits / ContractComplexity                        (3)
```

According to these formulas, simple interfaces that offer powerful functionality, by definition, have a much bigger depth than complex interfaces that offer little functionality. We'll call the former "deep" and the latter "shallow."[53] The reason for that is purely geometrical—if we visualize interfaces as rectangles, where their width represents the interface complexity and area represents the offered functionality, then deep interfaces look vertical, or you can say deep (see Figure 4.2).

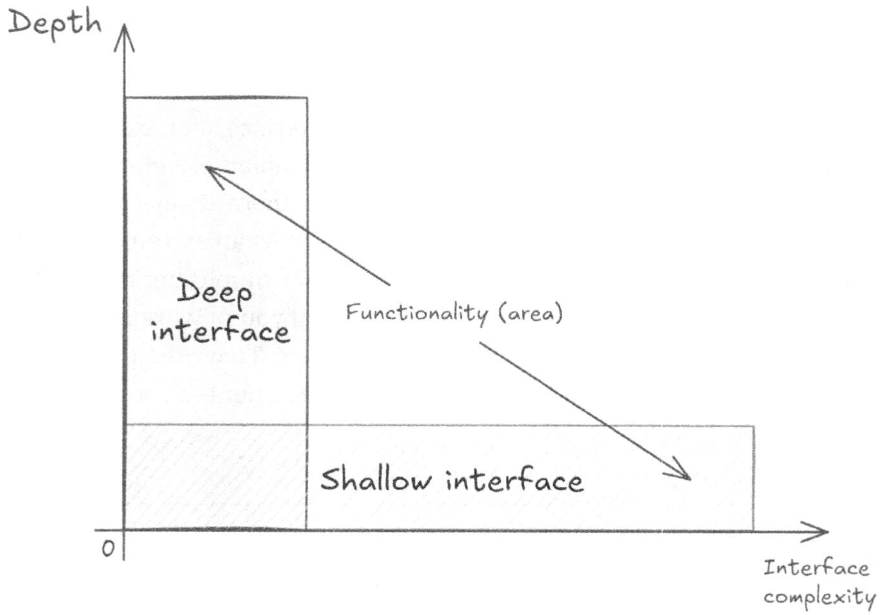

Figure 4.2. Deep and shallow interfaces.

---

[53] Following the terminology Ousterhout proposed for deep modules in *A Philosophy of Software Design* [Ousterhout18]. Yet note that I define depth a bit differently, with a focus solely on interfaces.

Dependencies with deep contracts provide you with a substantial value at the cost of adding a little bit of complexity. In contrast, dependencies with shallow interfaces provide little value at the cost of adding substantial complexity.

A Garbage Collector (GC) is an example of a module with an extremely deep interface. Its functionality scope is huge—with GC present, you don't need to bother with manual memory allocation anymore as GC automates it for you! Simultaneously, its interface for most use cases is virtually nonexistent. You don't need to make any function calls on program start or provide any configuration. It just works out of the box, which makes its interface complexity close to zero. So, according to formula (2), its depth approaches infinity.

Git is a module with a much shallower interface, because despite being super flexible and offering powerful functionality, its interface is notoriously complex. Finally, Listing 4.3 is an example of a module with an extremely shallow interface that is too complicated for the tiny functionality it provides.

```
static void addElementIfPositive(
    List<Integer> mutableList, int elementToAdd) {
  if (elementToAdd > 0) {
    mutableList.add(elementToAdd);
  }
}
```

Listing 4.3. Function with an extremely shallow interface.

Note that modules with both deep and shallow interfaces can have either a large or a small implementation (see Figure 4.4). The definition of depth says nothing about the implementation, implementation size, or implementation complexity. For convenience, we'll call modules with deep interfaces deep modules, but it's important to remember that depth is a property of an interface that has nothing to do with implementation.

Pause and Think

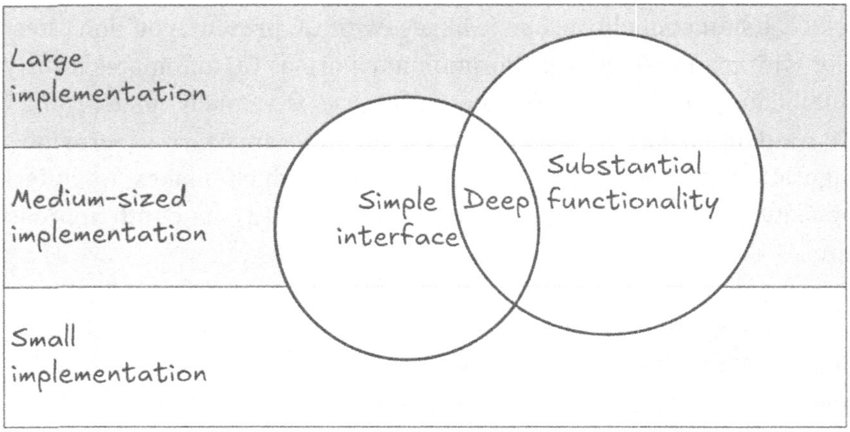

Figure 4.4. Both deep and shallow modules can have either a large or a small implementation (Venn diagram).

Although most shallow interfaces are not as bad as the one from Listing 4.3, none of them offer you the best value for money—either their contract is not simple enough or the functionality provided is too small, or both. If the main problem is complexity, this will lead to all sorts of issues we've already discussed above (see Figure 4.1). However, if the interface is simple but still shallow because of small functionality, then such an interface merely shifts complexity onto its users.

For some mysterious reason, we developers love talking about providing "building blocks" for other developers and other teams. This may be fine if we're building a public API or if there are too many use cases to analyze them all. But so often, the "building blocks"-style talk serves as a cover for not understanding well the use cases for the modules we're building. If somebody says, "Well, you can do whatever you want with this piece of software," chances are they have no idea of any truly useful applications! They may be lucky, and their building blocks may actually happen to be useful, but even then, it's highly unlikely that they would be *easy* to use.

> To use modules with simple but shallow interfaces, one needs not only to learn the interfaces themselves but also master their usage patterns.

At Google, our team was once working on a brand-new API, but we suffered from the problem described above—we hardly understood the use cases of other teams who would actually use that API. So, we ended up wasting multiple weeks building a granular CRUD API that turned out to be partially useless and partially hard to use. Luckily, we quickly spotted the problem and switched to building a truly product-centered API that was *way* deeper than a bunch of basic CRUD operations. The lesson here is rather universal: to use modules with simple but shallow interfaces, one needs not only to learn the interfaces themselves but also master their usage patterns. Even if doable, it's rarely productive.

4.3. Deep or small?

Even though too little functionality is not a rare problem, more often, shallow interfaces are shallow because of their complexity. But why would anybody ever overcomplicate an interface? True, often that's simply lack of skill or negligence, but that's not always the case. The most popular reason for consciously overcomplicating interfaces is to keep modules small. "Your functions shouldn't be longer than 50 lines of code," and "any class should have only one reason to change"—if you're not new to programming, I'm sure you've heard such recommendations (and probably felt guilty for not following them). Yet it's hardly ever possible to keep all the interfaces simple if all your modules are small.

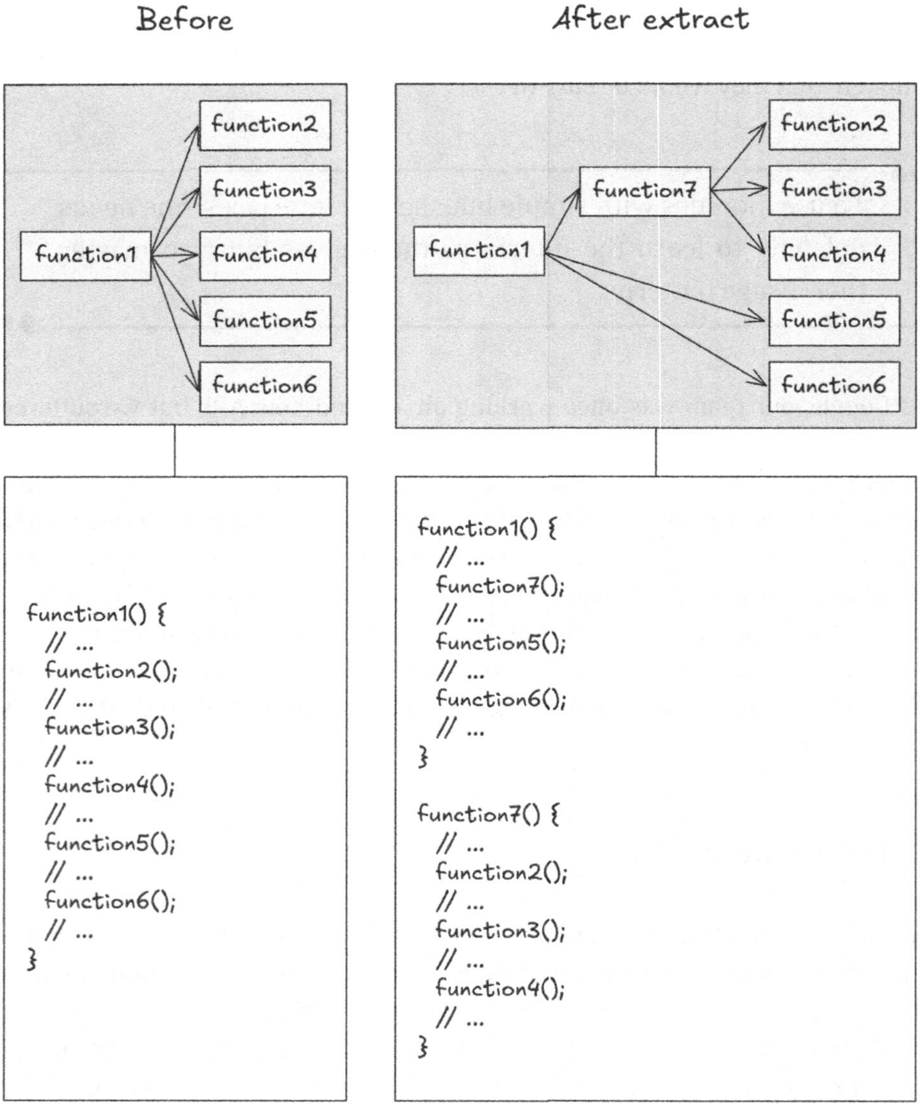

Figure 4.5. Break-up of a large function.

Consider having a big module, say a big function like `function1` on the left side of Figure 4.5. According to the small modules doctrine,[54] we need to break it up into smaller functions. So let's extract a middleman—`function7`. If it has a simple interface, then it's likely a great refactoring, but if not, we've just created a shallow module! Although such an artificial middleman makes the original module's (`function1`) implementation shorter, it hardly simplifies anything! The resulting modules are tightly coupled, hard to understand in isolation, and even force readers to read implementations rather than interfaces, undermining the whole idea of modular design.

There is nothing inherently wrong with the idea of keeping modules small. Small functions, small classes, small packages, small microservices—all of that sounds great! They are typically more readable than larger modules, being more cohesive and having a smaller number of dependencies (see Figure 5.4, for example). But as with most good ideas, you can take it too far. As we've seen above, when small size becomes a doctrine, it promotes the creation of shallow interfaces. Moreover, modules that are *too* small also tend to be dysfunctional—they tend just to delegate functionality implementation to other modules. We'll discuss dysfunctional layers in detail in the next chapter (see Section 5.5), but long story short, they may be the source of annoyingly frequent change amplifications. Consider the `smallDeepFoo` function from Listing 4.6, for example. It delegates most of the work to `deepFoo`, so *the added value is small*, and the change amplification risks are high.

```
void smallDeepFoo(params) {
  logger.info("Executing foo");
  publishFooCalledEvent(...);
  deepFoo(params);
}
```

Listing 4.6. Deep but dysfunctional small function.

[54] That's how I'll refer to the unhealthy obsession with the idea of small modules.

Finally, the last potential problem with the small modules doctrine is the total number of interfaces in your system. If there are too many, the *unfamiliarity* complexity cause comes into play as too many interfaces now compete with each other for being widely known and familiar.[55] However, it's worth noting that this problem is mostly mitigated by minimizing the number of *public* modules. If you have a small number of public and company-wide APIs; and a reasonable number of public classes and public methods in those classes, then the total number of modules in your system doesn't affect its overall complexity much. It's mostly public interfaces that compete with each other for being widely known and familiar.

Pause and Think

Avoid bloated interfaces

It's easy to fall into the trap of creating bloated interfaces with slight variations of the same method, such as `GetCustomerById`, `GetCustomerByEmail`, and `GetCustomerByPhoneNumber`. Using a single, flexible `GetCustomer` method with additional parameters is often a better approach. Yet, keep an eye on the number of parameters and their complexity to ensure it doesn't spiral out of control.

> Aim at building all your modules deep, and try to keep them small as long as it makes them more readable and not dysfunctional.

To summarize, smaller modules have one huge benefit—they tend to be more readable, but also, two major caveats to avoid: being shallow and being dysfunctional. Putting this all together, I would say aim at building all your modules deep, and try to keep them small as long as it makes them more

[55] That's a problem similar to the one discussed in Section 7.4 (see Figure 7.1).

readable and not dysfunctional. But why in this order? Why does the interface depth factor take precedence over a small implementation size? To answer this question, let's take a step back and take a look at what we've learned so far. First, we've discussed tons of reasons for the interface-first mindset at the beginning of this chapter. Then, in the previous section, we saw how massively important it is to keep interfaces not only easy to read but also easy to use. Combining these two truths typically[56] translates into a single statement: interfaces *must* be deep. Moreover, if the interface is not simple (=> not deep, see (2)), it's almost guaranteed that the implementation won't be readable as well, regardless of the module's size. That's why as a rule of thumb:

1. Aim at building all your modules deep.
2. Try to keep them small as long as it makes them more readable and not dysfunctional.

Pause and Think

Don't miss an opportunity to extract a module that is deep enough

In larger modules, constantly watch for pieces of code that are organically extractable as smaller deep modules, and extract them proactively even if they will never be reused. That not only improves readability but also prevents potential DRY principle violations (we'll talk about it more in the next chapter):

```
void foo() {
  // ...
  var y = calculateY(x);
}

// this deep method is called exactly once:
private int calculateY(int x) {
  // ...
}
```

[56] With a few rare exceptions mentioned in the previous section.

Finally, let's wrap it up by taking a quick look at Table 4.7, which is a trade-off table with a summary of the pros and cons of deep, shallow, and small modules. We'll discuss such color-coded[57] trade-off tables in detail in Chapter 8: "Trade-off Analysis and Design Docs," but the idea behind them is fairly simple. Let's say you want to choose among several options: first, you arrange them as rows in the table and put the factors affecting the decision-making as columns. Then you basically want to choose the best[58] row. Instead of focusing on nuances here, let's just make a couple of obvious observations from Table 4.7:

- Deep modules are the clear winners in the simplicity contest (see row 1).[59]
- Interface simplicity is typically much more valuable than powerful functionality (see rows 2–4).[60]
- There is not much we can say about small modules in general, except that they are more likely to have a simple implementation (see the last row).

The best modules are those that have simple interfaces and provide powerful functionality. Small modules are great if their interfaces are deep enough. Yet, contrary to the small modules doctrine, the deep modules one sets you free to write large modules if decomposing them into smaller ones is problematic. It's a popular myth that small modules (like small methods and small classes) are inherently good. Deep interfaces are.

> It's a popular myth that small modules are inherently good. Deep interfaces are.

[57] In a book I have to use shades of gray, but normally you would use red/orange/yellow/green instead (and white for N/A and question marks). You can find examples of the original color-coded ones at `https://codehealthguardian.com`.
[58] Note that columns may have different priorities.
[59] If you can improve implementation readability by making them small, that's even better!
[60] Unless you heavily prioritize some of the desired properties over the others.

	Easy-to-read interface	Easy-to-read implementation	Easy to use	Causes frequent change amplifications	Increases the total number of modules	Simplifies testing
Deep	Yes	May be long	Yes	No	Maybe	May be not too granular
Shallow— complicated interface	No	Rarely	No	Often	Maybe	No
Shallow— small functionality under a simple interface	Yes	Yes	Maybe	Often	Yes	Makes granular testing possible
Shallow— small functionality under a complicated interface	No	Rarely	No	Often	Yes	No
Small	?	Likely	?	Maybe	Maybe	?

Note: White is good, dark gray is bad. Question marks indicate no estimates.

Table 4.7. Trade-offs of deep, shallow, and small modules.

4.4. The power of defaults

Now, let's take a look at several nontrivial interface simplification techniques. In Section 4.2, I mentioned the Garbage Collector as an example of a super deep module. Although it may have tons of parameters for fine-tuning, thanks to defaults, most developers don't even need to know these exist. Just imagine writing a "Hello, World!" in Java involved something like

```
-XX:+UseG1GC
-XX:MaxGCPauseMillis=200
-XX:ParallelGCThreads=20
-XX:ConcGCThreads=5
-XX:InitiatingHeapOccupancyPercent=70
```

Fortunately, there are reasonable defaults for each of these parameters that work great in most cases and fine in others.

Yet there is one major caveat that substantially limits the applicability of this interface simplification technique—defaults shouldn't be surprising to anyone! Otherwise, they increase obscurity more than they decrease interface complexity (see Figure 4.8).

> Defaults shouldn't be surprising to anyone. If in doubt, avoid defaults and make it explicit.

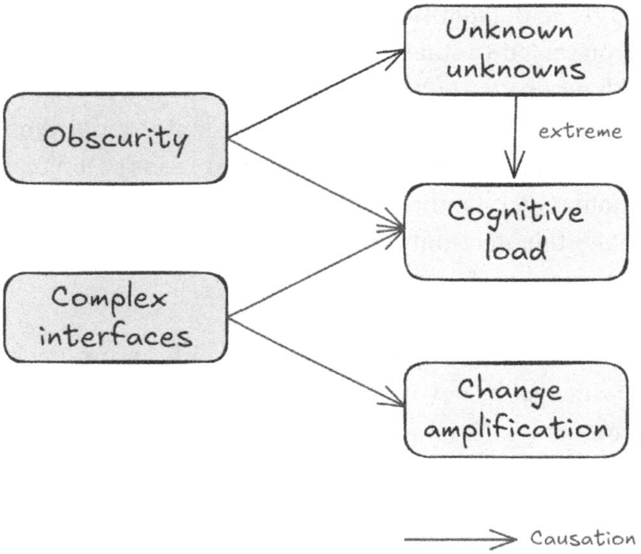

Figure 4.8. Defaults-related complexity causes and problems (also see Figure 2.3).

For example, let's say you have an API that reads from the database (DB). As your product becomes popular and the load increases, you start hitting the limits of your DB scalability. You search for a solution and bump into a popular quick fix—just add a read replica! Yet read replicas may return stale (read: wrong) data, so you decide to introduce a new optional boolean HTTP request parameter `useStaleDbRead`,[61] thus putting callers in control and letting them choose if it's okay to read from a read replica when processing that particular request. At first glance, it looks like a good backward-compatible API change that puts the callers in control. But the question is, what is the default value? If it's `useStaleDbRead=false`, unless you spend some time to go over the high-QPS calls to that endpoint and manually specify `useStaleDbRead=true` for some of them, it's unlikely that the original scalability issue would be addressed.[62] So there is a strong temptation to silently go forward with the

[61] Or, even better, an enum parameter such as `priority = (latency|accuracy)`.
[62] You may be lucky, and it might be handled by someone else.

`useStaleDbRead=true` default. However, even if documented, that's obscure. The problem is you can't do a stale read without the caller's explicit permission to do so—either have `useStaleDbRead=false` as a default or create a separate method for stale reads and name it accordingly (e.g., `ListUserTeamsStaleRead` in addition to `ListUserTeams`). Don't compromise correctness without understanding the consequences first,[63] and even more so, don't silently make this decision for others. Above everything else, defaults shouldn't be surprising!

Finally, there is one more factor to take into account. From my experience, most software issues stem not from poor decisions, but from indecision—simply when something falls into a blind spot. While reasonable defaults is a powerful simplification technique, often, you'd want to make callers think about a particular aspect of what they're about to do. They well may not even realize that they should have made a decision unless you literally force them to. For example, in the past, logging libraries commonly used to have a `log(message)` function without an explicit logging level (`INFO`, `WARN`, `ERROR`), usually using `INFO` as a default. Yes, it is a reasonable default that would hardly be surprising to anyone, but is it a good enough reason to have it? Probably not. It's better to nudge developers to think about the right logging level whenever they log something. That's what most libraries do today, offering only `logger.log(message, level)`, `logger.info(msg)`, `logger.warn(msg)`, and alike—functions with an explicit logging level.

Defaults are a powerful interface-simplification technique, but use them carefully. "Defaults shouldn't be surprising" is a necessary but not always a sufficient condition for introducing a default. If in doubt, avoid defaults and make it explicit.

---

[63] Unless there is a real risk of the system being down, of course. In such cases, compromising correctness should be viewed as a temporary quick fix.

4.5. Somewhat general-purpose interfaces

When discussing the building blocks mentality, we saw how *not being mindful* of user needs can lead to interface shallowness. But, interestingly, you can end up there if you *overfocus* on the problems at hand as well. I know, that may sound suspicious—is it really possible to over-prioritize the current requirements? Doesn't it contradict the famous YAGNI[64] principle? Well, it does a little bit, so let me explain.

A concept of a "somewhat general-purpose" interface was introduced by Dr. Ousterhout in his book *A Philosophy of Software Design* [Ousterhout18]. It starts with an observation that general-purpose interfaces[65] are often simpler than their special-purpose counterparts. They're almost always more readable, and sometimes, they may even be easier *to use* as well, especially if they're not *too* general-purpose. I know that sounds a bit abstract, so let's take a look at a slightly modified example from the same book. Let's say we're implementing a text editor and in the UI code we delegate making the file changes to a lower-level `Document` class:

```
/** Deletes a symbol to the left of the cursor. */
void handleBackspace() {
  // ...
  // Delegate to low-level document class object
  document.backspace(cursor);
}

/** Deletes a symbol to the right of the cursor. */
void handleDelete() {
  // ...
  document.delete(cursor);
}
```

---

[64] Short for "You Aren't Gonna Need It," a principle that arose from extreme programming (XP) according to which a programmer should not add functionality until deemed necessary.

[65] Those addressing a broader range of problems, unlike special-purpose ones.

First of all, have you noticed a code smell here? It's the use of the "cursor" abstraction in the Document class—a storage layer class that works with the File System. Normally, UI abstractions shouldn't leak into the storage layer (see Section 5.5). After taking a closer look at the `delete` and `backspace` methods, it becomes obvious that they're doing the same thing. Yes, in UI, they are two different operations, but from the File System perspective, there is no difference. So let's replace those two with a single more general-purpose method:

```
interface Document {

  // ...

  void delete(Range<LineAndColumn> positionRange);
}
```

The new interface is not only more readable, but it also perfectly covers both use cases.[66] Whenever you're designing a nontrivial interface, ask yourself whether making it a little bit more general-purpose would simplify it. If that's the case, then writing a longer implementation and additional test scenarios may be totally justified.

4.6. Avoidable edge cases

To get an idea of how annoying edge cases are, you don't even need to have any real work experience—all you need to do is try to prepare for a technical interview. While some edge cases are inherent to the nature of software and the complexity of real-world systems, some of them are simply the result of a bad interface design. I'm sure most of us debugged getting `null` from a

[66] It also happens to be more future-proof. Once we get to the selection deletion functionality, we'll just reuse the same `delete(positionRange)` method. But be careful with such arguments—make sure you're not overcomplicating the interface for some theoretical future use cases. That would be a truly bad YAGNI principle violation.

findUsers function invocation (instead of getting an empty collection). Actually, whenever the return type is truly nullable, and even if properly documented, that's an edge case that callers need to be aware of and process differently. But it's not only nullable return types—errors that may be returned or exceptions that may be thrown are also a part of the interface and in fact are also edge cases. Consider the following two factory methods:

```
/**
 * @throws IllegalArgumentException if level is the last one.
 */
static PointLevel of(Level level, double value) {
  // ...
}

/**
 * @throws IllegalArgumentException if level is not the last one.
 */
static PointLevel ofLastLevel(Level level, double value) {
  // ...
}
```

The last level (whatever it means) is an edge case that won't be processed by PointLevel.of. That means callers would need to do something like this:

```
var pointLevel =
    level.isLast()
        ? PointLevel.ofLastLevel(level, value)
        : PointLevel.of(level, value);
```

It's really hard to justify the existence of the "last level" edge case in this interface. Instead, it should have been something like this:

```
static PointLevel of(Level level, double value) {
  if (level.isLast()) {
    return ofLastLevel(level, value);
  }
  // ...
}

private static PointLevel ofLastLevel(Level level, double v) {
  // ...
}
```

Unfortunately, some interfaces even allow for *self-contradictory* parameter values, like in the following snippet:

```
Report generateReport(
    String dataFormat,
    boolean includeImages,
    boolean includeTables) {
  if ("JSON".equals(dataFormat) && includeImages) {
    throw new IllegalArgumentException(
        "JSON files can't have images.");
  }
  if ("CSV".equals(dataFormat) && includeTables) {
    throw new IllegalArgumentException(
        "CSV files can't have tables.");
  }
  // ...
}
```

That's an even worse type of edge case. Here, the callers must be aware of the internal constraints between the parameters. That's totally avoidable—just have a separate function for each of the supported formats: `generateJsonReport`, `generateCsvReport`, `generatePdfReport`, etc.

Finally, I can't stop myself from mentioning a classic example of throwing an exception when deleting nothing:

```
@Nullable
Selection selection;

/** @throws IllegalStateException if nothing is selected. */
void onSelectionDelete() {
  if (selection == null) {
    throw new IllegalStateException("Nothing is selected");
  }
  // ...
}
```

An obvious alternative would be just to do nothing instead:

```
/** ... If nothing is selected - it's NOP. */
boolean onSelectionDelete() {
  if (selection == null || selection.isEmpty()) {
    return false;
  }
  // ...
}
```

You can actually think of avoiding edge cases in terms of making the interfaces more general-purpose, very much in line with the principle we discussed in the previous section.

Pause and Think

There are enough edge cases inherent to the nature of software and the complexity of real-world systems for us to add even more of them to our codebases. If an edge case is avoidable, please do avoid it, especially in interfaces.

Summary

Simple interfaces are your top priority as a Code Health Guardian. "Simple" here means not only easy to read but also easy to use, which typically means deep. Aim at building all your modules deep, and try to keep them small as long as it makes them more readable and not dysfunctional. At the same time, feel free to write large modules if decomposing them into smaller ones is problematic.

In this chapter, we've discussed three particular interface simplification techniques, but these are not the only ones covered in this book. When discussing code health principles and techniques in the following chapters, we'll focus more on how they are applied to interfaces rather than implementations. That's the interface-first mindset in practice.

As a reminder, chapters 3–10 are mostly orthogonal to each other and greatly differ in complexity, so feel free to read them in an order that's most meaningful to you, or even skip a particular chapter.

Chapter 5

INDEPENDENCE QUEST

Understanding a function that makes 20 method calls across 10 different classes may be challenging. It's hard to reason about code with too many dependencies. Yes, there are trivial ones, such as `Math.max(x, y)` calls that can hardly complicate anything, but that's more of an exception. Most dependencies *do* count toward our brains' quota, whether these are local variables, parameters, fields, invariants, or function calls. So, what about making our code more independent? That's the idea explored in this chapter.

> The "too many dependencies" problem is relevant at every level: a *function* may have too many dependencies, or a *class*, or even a *microservice*, or some *high-level design*.

As a reminder, for the purposes of this book, a dependency exists when a given piece of code cannot be understood or modified in isolation, which is commonly referred to as coupling. In Chapter 2, when talking about the "too many dependencies" complexity cause, we compared our brain's working memory to a limited number of CPU registers. However, this analogy isn't flawless as it might give an impression that this complexity cause is relevant to individual statements only. While a single logical transition involving multiple facts can definitely be complex, the "too many dependencies" problem is relevant at every level: a *function* may have too many dependencies, or a *class*, or even a *microservice*, or some *high-level design*. In this chapter, we're going to explore the idea of having fewer of them—fewer dependencies, or less coupling, if you will.

This chapter opens with a section on knowledge duplication, which is basically coupling in its extreme form. Then we'll talk about code discoverability, using the right abstractions, and anemic domain models, wrapping it up with a tempting idea of introducing an extra level of indirection.

5.1. Don't Repeat Yourself

If our goal is having fewer dependencies, what about getting rid of one of them by just inlining it? Let's say the inlined code doesn't have any nontrivial dependencies of its own (see Figure 5.1). What do you think of such refactoring?

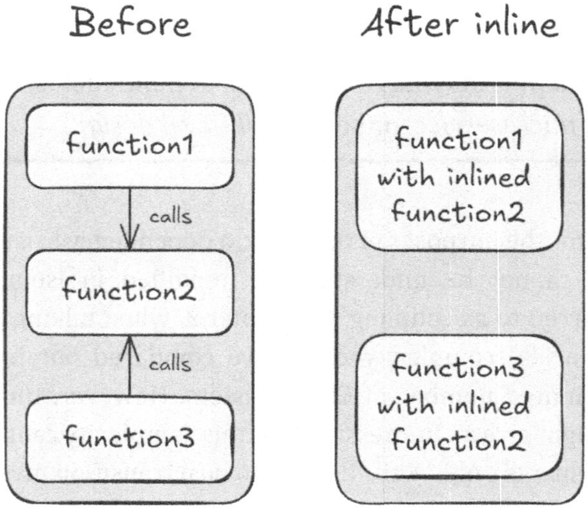

Figure 5.1. Function inline.

Looking at our definition of dependency, it's easy to spot the problem—the dependency is still there! If we need to change the implementation of function 2, now we need to change it in both places. Actually, when inlined, it's even worse than just a dependency. It's an *obscure* dependency now—a dependency that causes not only change amplification problems, but unknown unknowns as well. Just as a reminder, in Chapter 2, we ranked the three complexity problems, change amplification being the least critical and unknown unknowns the most critical of them. Yes, making changes in several places is definitely an overhead, but it's rarely a big deal—unless there are *too many* of them, of course.[67] The worst problem with duplicates is definitely not change amplifications. What if the supposed duplicates are out of sync with each other? Let's say if some changes were applied to one of the duplicates but not to the other? Giving a computer two contradictory pieces of knowledge rarely ends well. That's what made the authors of *The Pragmatic Programmer* formulate the famous Don't Repeat Yourself (DRY) principle, which I believe is one of the most important principles to follow to keep your codebase healthy:

> *Every piece of knowledge must have a single, unambiguous, authoritative representation within a system.* [HT19]

"Duplication" and "too many dependencies" complexity causes
You can think of a knowledge duplicate as a dependency taken to the extreme. So one could argue that the "too many dependencies" complexity cause somewhat covers duplication as well (see Figure 2.3). However, a knowledge duplicate is a very special case of a dependency that has quite atypical consequences and requires a special approach. Therefore, I believe it deserves its own place among the seven causes of complexity.

Although the DRY principle is so massively important, over the last few decades, there was quite a lot of confusion and controversy around it. Let's take a quick look at the code below. Can you spot any duplicates?

---

[67] In *The Pragmatic Programmer* [HT19], the authors share a story of an audit company finding thousands of implementations of SSN (a nine-digit number with a couple of hyphens) validation logic in a U.S. governmental codebase. So, when the U.S. government runs out of available SSNs, which is likely to happen in this century, that would be a huge refactoring for them to do. Well, not exactly for them.

```
public class Vector {
  Point start;
  Point end;
  double length;

  // ...
}
```

The key observation here is that the DRY principle is about *knowledge* duplicates, not about *code* duplicates. Notice that the definition above says "piece of knowledge," not "piece of code." In the vector example above, the knowledge about its length is encoded in two places: explicitly in the `length` field, and implicitly in the vector's coordinates. Don't express the same thing twice, even in two totally different ways. Not every knowledge duplicate is a code duplicate.

Actually, the reverse is true as well—not every code duplicate is a knowledge duplicate. For example, if two fields happen to have the same constraints on them, having two copies of code doing the same constraints validation is not a violation of the DRY principle. Consider the following code:

```
verify(quantity != null);
verify(quantity > 0);

verify(price != null);
verify(price > 0);
```

It's not hard to imagine `price=0` becoming a valid value in the future, which has nothing to do with the constraints on `quantity`. These checks should exist and evolve independently of each other. It's important to remember that DRY is about knowledge duplication, not about code duplication.

Pause and Think

DRY and shallow interfaces

In the previous chapter, we learned how crucial it is for interfaces to be deep. Yet as far as I'm aware, there is one solid reason for having a module with a shallow interface—avoiding duplicates. As we'll see in the following, it's not always the best option, but it's definitely an option. Creating shallow interfaces used in several places to avoid knowledge duplication is typically justified.

That said, although I'm a strong advocate of the DRY principle, I believe the way it is defined in *The Pragmatic Programmer* has one problem—it's overly restrictive. It says that "every piece of knowledge *must* have a single, unambiguous, authoritative representation within a system." But is it really a must? Let's go back to the first principles and remind ourselves of the practical problems with duplicates. That's where differentiating between the three complexity problems comes to help. Duplicates will always cause change amplifications, which is unavoidable, but the good news is we can easily avoid the problem of unknown unknowns. The trick is to leave a comment in each of the duplicates saying where else that knowledge is duplicated, or in other words, leave bidirectional[68] pointing comments (see Listing 5.2). At Google, we even had a pre-commit automation in place that would notify you if you changed one of the duplicates but not the other in your PR.[69] That's a handy tool, but in practice, most of the value comes from the comments themselves—it's nearly impossible to modify a duplicate without noticing the comment saying "please modify it in some other place as well."[70] This workaround enables you to violate the DRY principle without any serious consequences.

---

[68] Strictly speaking, we could call them multidirectional if referring to several duplicates.
[69] Internally called CL (Change List).
[70] The syntax from Listing 5.2 without automation may not be good enough for longer duplicates as it becomes easier to miss the surrounding comment when making a change. So, without the pre-commit automation, for big duplicates you would likely need a slightly different syntax—maybe just have the same comment every 20 lines, saying that this code block is duplicated.

```
// book.ts

// LINT.IfChange (someLabel)
// Code duplicate here.
// LINT.ThenChange(backend-
service/src/main/java/Book.java:anotherLabel)

// Book.java

// LINT.IfChange (anotherLabel)
// Code duplicate here.
// LINT.ThenChange(frontend-service/src/book.ts:someLabel)
```

Listing 5.2. Bidirectional pointing comments make duplication explicit.

Tests: Make them DAMP, not DRY

Code health standards for tests are very similar to those of the production code, with just a couple of differences. One of them is that it may be okay to violate the DRY principle in tests, especially if complying with DRY would lead to the creation of complex interfaces or deep call stacks. Clarity in tests is even more important than in production code. As we don't write tests for tests, they should be self-evidently correct. Yet another reason why violating the DRY principle is not a big deal in tests is that it rarely causes the problem of unknown unknowns, as tests are likely to fail if you forget to change any of the duplicates. Thus, instead of being DRY, tests should be DAMP, meaning they should promote "Descriptive And Meaningful Phrases" [WMW20]. They may be DRY as well, but that's just nice to have.

While saying that DRY is a must would be an overstatement, it's strongly recommended to avoid any knowledge duplicates—in your code, in your documentation, and in your data. Try to make all that DRY by default. However, whenever you need to violate the DRY principle for performance, reliability, or some other serious reasons, leave the bidirectional pointing comments there[71] and don't feel bad about it.

[71] When talking about data, obviously, this technique works for DB schema files only.

5.2. Code discoverability

An inherent challenge of the DRY principle is that developers often violate it unintentionally, creating duplicates without even realizing it. In fact, how do you know that some of the code you're writing doesn't already exist somewhere in the codebase? Mandatory code reviews definitely help with such knowledge sharing, but they're more like a safety net. The real solution is to make your code easily *discoverable*—to make it easy to find the existing functionality. If that's not the case, just give it enough time, see the most knowledgeable team members leave, and you will soon be lost in unintentional DRY violations. Interestingly, at Google, we had this problem in our three-year-old project, even with all the original authors still on board. For multiple reasons, it wasn't easy answering whether we already had a certain functionality somewhere in our codebase. It's no surprise then that we were regularly bumping into unintentional duplicates, as that's an inevitable consequence of bad code discoverability.

There are two key factors affecting code discoverability: structure and searchability (see Figure 5.3). Structure is about organizing code in some logical hierarchical manner, such as grouping code in functions, functions in classes, having those functions in some logical order within a class, and organizing classes into directories/packages. I know it's basic stuff, but having a clear code structure helps a lot with code discoverability, and on some projects this basic stuff is undervalued.

Yet even with all the structure in place, it's rarely possible to manually navigate to a single place where the code you're looking for may potentially live. That's why we use code search all the time, right? To search across the whole codebase, within the directory, or just within a file. Code searchability is better in codebases with the right abstractions, descriptive names, and all the necessary comments and documentation in place. That's where it gets really interesting—given that good code discoverability helps avoid unintentional DRY violations, we see again how healthy code helps keep it healthy, while complex code leads to even more complexity over time (see Figure 1.1).

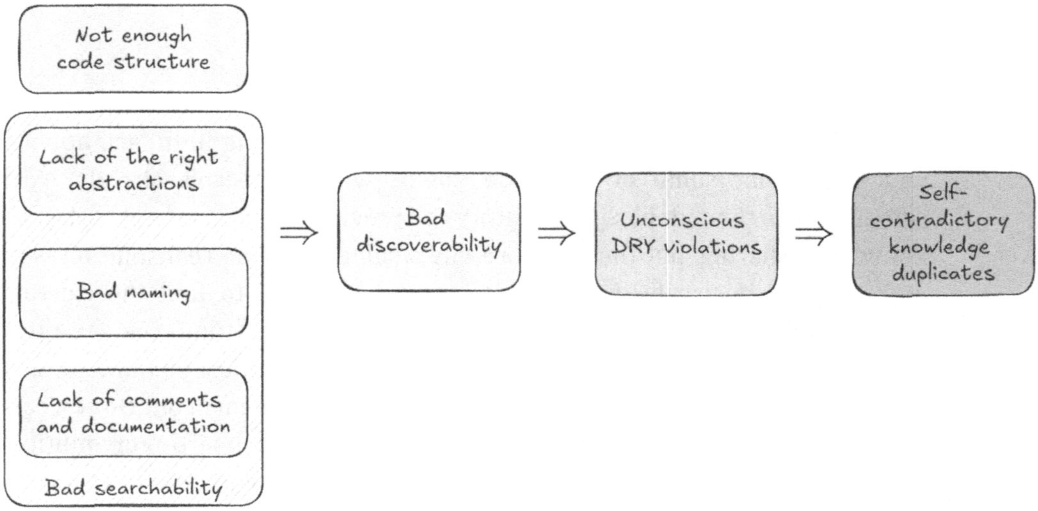

Figure 5.3. How bad code discoverability ruins systems.

> Healthy code helps keep it healthy, while complex code leads to even more complexity over time.

There is a famous quote from *Clean Code* that I think is still true today: “The ratio of time spent reading versus writing is well over 10 to 1. We are constantly reading old code as part of the effort to write new code” [Martin08]. Yet, I would say in some codebases a comparable amount of time is spent navigating the code—searching and skimming through it, just looking for something. The good news is that as AI-powered tools mature, in the very near future they’ll definitely simplify the process of finding existing code for us, and will even proactively help us with spotting knowledge duplicates. Nevertheless, good code structure and searchability will continue to play an important role in code discoverability, as they’ll definitely affect the quality of AI assistance.

5.3. The right abstractions

Now let's go back to our challenge of trying to reduce the number of dependencies. Even though I've opened this chapter with an inline example (see Figure 5.1), it's exactly the opposite that can actually help us minimize their number. Let's take another look at Figure 4.5—the function extraction example from the previous chapter. But let's do it from a slightly different angle now, with a focus on the number of dependencies. After the extraction, both `function1` and `function7` technically have fewer dependencies, yet whether or not it reduces the overall complexity ultimately depends on the *interface* of the extracted `function7`. If its interface is clear, then it's a great refactoring, but if it's untrue or complex, that would usually outweigh the benefits of technically having fewer dependencies (see Figure 2.3). As already mentioned in the previous chapter, if the interface is unclear, the tight coupling that we've tried to avoid is de facto still there—to really understand what `function7` is doing, developers would need to read `function1`, and vice versa.

Exactly the same guiding principle of "interface simplicity" applies to classes and other types of modules. As a rule of thumb, if it's easy to name a module, it's likely worth having it, but if any name you can think of is confusing or obscure, then consider some other ways of decomposing your code into modules.[72] Listen to bad names as they may signal the weaknesses of the current design (see Section 7.7). More broadly, it's not only the name but overall simplicity of the interface that matters. Simple interfaces, and simple names in particular, are by far the best symptoms of a good decomposition into modules.

It's interesting to see how exploring the idea of having fewer dependencies leads us back to the familiar territory of simple interfaces. I think everyone would agree that finding the right abstractions is *the most* important interface simplification technique. And vice versa, simple interfaces are the best validation that you've found the right abstractions—they're the most reliable

---

[72] Merging it with some other module it's tightly coupled with may be a good starting point.

signal that you've got the decomposition into modules right. It's fascinating how different complexity causes such as "too many dependencies," "complex interfaces," and "obscurity" in naming intertwine together in the code decomposition challenge, which is definitely the hardest problem in managing code complexity.

Let's pause for a moment and take a look at Listings 5.4 and 5.5. Which version of the `createPizza` function do you think is more readable?

```
class Order {
  Pizza createPizza() {
    var pizza = Pizza.prepare(this);
    pizza.bake();
    pizza.box();
    return pizza;
  }
}

class Pizza {
  void bake() {
    var oven = Ovens.getFreeOven();
    oven.preheat();
    oven.bake(this);
  }

  static Pizza prepare(Order order) {
    var pizza = new Pizza(order.size, order.sauce, Cheese.REGULAR);
    pizza.addToppings(order.kind);
    return pizza;
  }
}
```

Listing 5.4. Create pizza with the proper abstractions.[73]

[73] Inspired by `https://testing.googleblog.com/2023/09/use-abstraction-to-improve-function.html`

```
Pizza createPizza() {
  var pizza = new Pizza(size, sauce, Cheese.REGULAR);
  switch (kind) {
    case MEAT -> pizza.toppings = Toppings.MEAT_TOPPINGS;
    case VEGETABLE -> pizza.toppings = Toppings.VEGETABLE_TOPPINGS;
  }
  var oven = Ovens.getFreeOven();
  while (oven.temperature < cookingTemperature) {
    sleep(checkOvenInterval);
    oven.measureTemperature();
  }
  oven.insert(pizza);
  sleep(cookTime);
  oven.remove(pizza);
  pizza.baked = true;
  var box = Boxes.create(/* ... */);
  // TODO: more code for boxing, slicing, etc.
}
```

Listing 5.5. Create pizza without the proper abstractions.[74]

You would probably agree that Listing 5.4 is simpler, but why? There are two major problems with the function from Listing 5.5:

- Too many objects, their fields, and dependencies between them
- Mixture of different levels of abstraction: low-level implementation details (e.g., how to heat the oven), intermediate-level functions (e.g., how to bake pizza), and high-level abstractions (e.g., preparing, baking, and boxing the pizza).

In contrast, Listing 5.4 leverages those abstractions to decompose code into smaller cohesive functions. Originally, the book's subtitle was *The art and science behind code simplicity*, and it's no coincidence that it had the word "art"

---

[74] Inspired by `https://testing.googleblog.com/2023/09/use-abstraction-to-improve-function.html`

in it—when looking for the right abstractions we're more artists than scientists. Abstractions that are native to your domain (such as "oven," "pizza," "prepare," "bake," and "box" in the example above) are usually a good starting point, but finding the right ones is typically a continuous process. You probably won't get it right the first time, and what "right" means will evolve together with your product, so be always on the lookout for better abstractions to make your modules more cohesive and less coupled.

To explore this topic further, consider reading a classic book by Eric Evans, *Domain-Driven Design* [Evans03]. However, don't expect too much from it—like with any other art, finding the right abstractions is more about your own practice in collaboration with others than about any theory.

5.4. Anemic domain models

Using an object-oriented language doesn't automatically make your code object oriented. Anemic domain models are a living proof of that—it's a design anti-pattern where domain objects lack behavior and contain data only. Listings 5.6 and 5.7 contrast anemic classes to those that properly encapsulate their behavior.

```
class AnemicRectangle {
  private final double width;
  private final double height;

  public AnemicRectangle(double width, double height) {
    this.width = width;
    this.height = height;
  }

  public double getWidth() {
    return width;
  }
```

```
  public double getHeight() {
    return height;
  }
}
```

Listing 5.6. Anemic class.

```
class Rectangle {
  private final double width;
  private final double height;

  public Rectangle(double width, double height) {
    checkArgument(width > 0);
    checkArgument(height > 0);
    this.width = width;
    this.height = height;
  }

  public double getWidth() {
    return width;
  }

  public double getHeight() {
    return height;
  }

  public double getArea() {
    return width * height;
  }

  // ...
}
```

Listing 5.7. Class that properly encapsulates its behavior.

It's not necessarily the hand-written (or AI-written) code that constitutes an anemic domain model: passing around generated classes that lack domain-specific behavior, like those generated by Protobuf[75] or Jooq,[76] is not much different. As Martin Fowler accurately notes, "The anemic domain model is really just a procedural style design."[77] That won't play out well for nontrivial applications as it leads to proliferation of `Util` and `Helper` classes and all kinds of static methods. Anemic domain models have not only poor code readability, but also poor discoverability, with all the ensuing consequences discussed earlier in this chapter (see Figure 5.3).

Consider the following code snippet registering a football analytics event:

```
match.getPlayerPresence(shotEvent.playerId)
      .getPresenceInterval(shotEvent.timestamp)
      .registerEvent(shotEvent);
```

Such chains of method calls also indicate a mild form of anemic modeling—the logic of registering an event itself is well-encapsulated, but that's not nearly enough. Instead, it should have looked like this:

```
match.registerEvent(shotEvent);
```

Getting rid of chains of method calls is a great opportunity to have fewer dependencies and evolve your domain model. When looking for the right home for a piece of code, ask yourself, "Where would other developers expect the code I'm writing now to live?" Combined with a continuous search for the right abstractions, that's the way to grow a rich domain model—to have the right abstractions with properly encapsulated behavior. I think those of us who were lucky to experience working with truly rich domain models won't settle for less again.

---

[75] https://protobuf.dev/
[76] https://www.jooq.org/
[77] https://martinfowler.com/bliki/AnemicDomainModel.html

Where does that method belong?

Consider the following method:

```
class ManualDataProviderController {

  // ...

  private boolean isDataDerivedFromGps(
      Space space, EventType eventType) {
    return eventType.getPrototype() == null
        || space.getIntegrations().stream()
          .flatMap(i -> i.getAbstractEventTypes().stream())
          .anyMatch(aet -> aet.equals(
              eventType.getPrototype().getAbstractEventType()));
  }
}
```

The first important observation here is that it can be a static method, as it doesn't use any of the fields or methods of the enclosing class. This means that we can make it a nonstatic method of one of its parameters. Given that `eventType` has a `space` field, we end up with an organic `eventType.isDataDerivedFromGps()` method instead of the above:

```
class EventType {

  // ...

  public boolean isDataDerivedFromGps() {
    if (getPrototype() == null) {
      return false;
    }
    var abstractEventType = getPrototype().getAbstractEventType();
    return space.getIntegrations().stream()
        .flatMap(i -> i.getAbstractEventTypes().stream())
        .anyMatch(aet -> aet.equals(abstractEventType));
  }
}
```

Notice that we not only got rid of the two extra parameters, but most importantly, we found the right home for this method, making it easily findable and preventing future DRY principle violations.

5.5. Dysfunctional layers

The fundamental theorem of software engineering states that "we can solve any problem by introducing an extra level of indirection,"[78] which is often expanded by the humorous clause, "except for the problem of too many levels of indirection." The intuition behind this humorous clause is that additional levels of indirection build up chains of dependencies, and "too many dependencies" is not only a problem for individual modules in isolation but for such chains as well, causing cognitive load and change amplification problems.

Pause and Think

There are many legitimate use cases for layers in software design. In fact, all modern software systems are composed of layers. There are even widely known layer-based design patterns that we work with on a daily basis, such as facade, MVC, and three-tier architecture. However, given that an extra layer can easily become a burden, it's vital to clearly articulate its purpose: which responsibilities would that layer *consistently* take, and how would that help the rest of the system?

At first glance, adding an extra layer by itself may look deceptively strategic and even scientific. Yet, the true science and good strategy are more about clarity than about sophisticated constructs. The main problem is that an extra level of indirection may be (almost) dysfunctional. You can spot it by seeing similar or even identical interfaces in adjacent layers, implementation simply delegating to the next layer:

[78] In his Turing Award Lecture, Lampson attributed this saying to David J. Wheeler. `https://bwlampson.site/Slides/TuringLecture.htm`

```
private Presence getPresence(Unit unit) {
  // ...
}

public Map<EventType, Lod> getLevelOfDetailByEventType(
    Unit unit) {
  return getPresence(unit)
      .getLevelOfDetailByEventType();
}

public Lod getLevelOfDetail(Unit unit, EventType eventType) {
  return getPresence(unit)
      .getLevelOfDetail(eventType);
}

public void setEventTypeLevelOfDetailForUnit(
    Unit unit, EventType eventType, Lod levelOfDetail) {
  getPresence(unit)
      .setEventTypeLevelOfDetail(eventType, levelOfDetail);
}
```

That obviously leads to constant change amplifications and hardly adds any value.

> Given that an extra layer can easily become a burden, it's vital to clearly articulate its purpose: which responsibilities would that layer consistently take, and how would that help the rest of the system?

And even if interfaces of adjacent layers are different but use the same *abstractions*, that may *still* be a code smell. For example, there is a widespread anti-pattern in Java of leaking HTTP request objects into an application layer from a Servlet[79] (the code snippet below is a little bit of an exaggeration):

```
class UserServlet extends HttpServlet {

  public void doGet(
      HttpServletRequest req, HttpServletResponse res) {
    // ...

    validateRetryConstraints(req, retryNumber);
    var user = userManager.getUser(req);

    // ...
  }
}
```

A proper behavior would be to unpack the request and pass down only the necessary pieces of data. Yet another widespread anti-pattern is doing the same thing, but in the other direction—when JSON or Protobuf[80] data from the storage layer leaks into the business logic.[81] It's better to avoid such information leakage in both directions (see Figure 5.8), as it unnecessarily complicates the application layer.

Pause and Think

[79] Servlet in Java is an entry point for HTTP requests processing.

[80] `https://protobuf.dev/`

[81] The best practice is to convert that data into normal objects as soon as possible. That not only guarantees low coupling, but also makes it possible to benefit from OOP in the business logic (domain) layer.

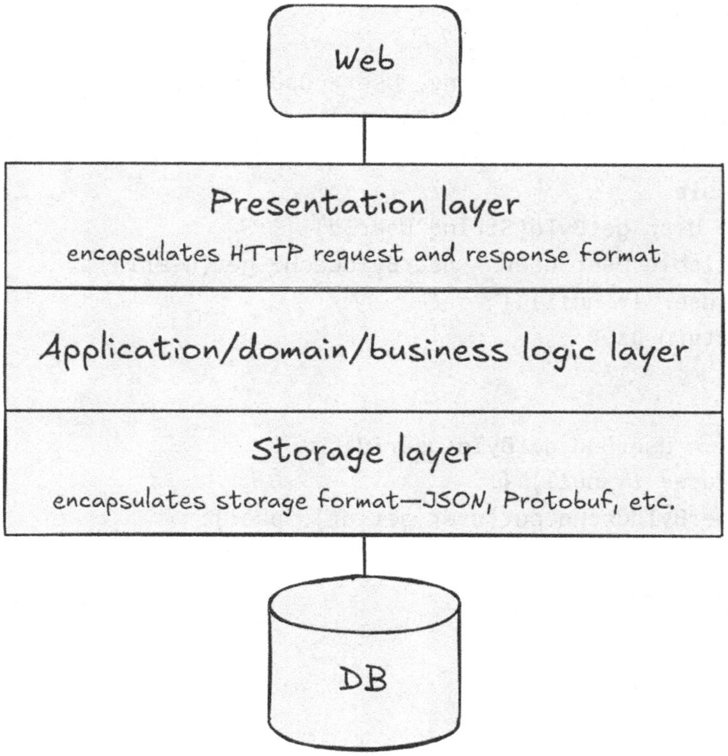

Figure 5.8. Three-tier architecture layers and the corresponding abstractions.

Using the same abstractions in adjacent layers can be a code smell—that may signal an information leakage unnecessarily complicating the adjacent layer. However, the fact that layers share the same abstractions or even have exactly the same interfaces doesn't *necessarily* mean one of them is dysfunctional. For example, we could have a caching layer on top of the storage layer with exactly the same interface:

```
public class UserService {

  private final Cache<String, User> userByIdCache;
  private final UserDao userDao;

  @Nullable
  public User getById(String userId) {
    @Nullable User user = userByIdCache.get(userId);
    if (user != null) {
      return user;
    }

    user = userDao.getById(userId);
    if (user != null) {
      userByIdCache.put(user.getId(), user);
    }
    return user;
  }

  // ...
}
```

What makes a layer dysfunctional is a lack of responsibilities that it would consistently take to unload the rest of the system. As long as a layer has a clear substantial set of responsibilities, the complexities of an additional level of indirection are often justified.

Summary

Given that dependencies are an indispensable part of any code, having less of them is tricky but doable. One of the least controversial and most important principles here is avoiding knowledge duplicates, which are essentially dependencies taken to the extreme. Don't violate the DRY principle without a good reason, and if violating it, use bidirectional pointing comments to mitigate the problem of unknown unknowns. To avoid *unintentional* duplicates, make sure your code is easily discoverable. Finding the right home for your code is an important part of that—build a rich domain model based on the right abstractions with a properly encapsulated behavior. Finally, avoid dysfunctional layers, that is, those that do not take any substantial responsibilities and mostly just delegate. Following these principles will help you have exactly as many dependencies as needed.

Chapter 6

SEMI-FUNCTIONAL PROGRAMMING

Perhaps to the disappointment of some and the relief of others, this chapter is not about functors, monads, Haskell, or Scala. Instead, it's about what I've learned as a Google software engineer writing in Java.[82] Even if you've never written a line of code in functional style, I bet you've heard about functional programming (FP) from your friends or colleagues. But do you remember exactly from whom? If it was one of your smartest friends, you're not alone. While the popularity of FP among the brightest minds stems from its academic roots, even more fundamentally, I believe it's a passion to understand code that attracts so many of them to FP. I know it may sound counterintuitive, because to many of us, even good functional code often looks unfamiliar and confusing. However, simplicity not only precedes complexity but also follows it. Although FP is more complex than mainstream imperative programming at first glance, it paves the way to more understandable code. Just bear with me for a moment, and I'll explain how.

When we say we understand something, we're often wrong. Even though as programmers we experience the veracity of this statement more often than other professionals, still, it may take us years to fully comprehend how true it is. We can spend many years believing that we understand the programs we're working on (or at least we're almost there),[83] only to wake up one morning and finally realize how fundamentally complicated programming really is [Barr18]. Yet, at this point, it may be easier to accept the status quo than to challenge it.

---

[82] Moreover, when I joined, it was Java 6 (no streams support); only later on we upgraded to Java 8.

[83] Especially, if we don't make it easy for our users to report bugs.

But not for those brilliant folks—they are something different (maybe you even find them a bit odd). They are extremely curious, and you can say even crazy about understanding things.[84] And I think the reason why so many of them are attracted to FP is that, in the FP universe, they feel they can reason about their own creations and not be wrong, which is a great value proposition, especially in the context of big complex software systems. Don't get me wrong, FP doesn't make the Herculean task of understanding them simple, but it does make it more tractable.

> More complex than mainstream imperative programming at first glance, functional programming paves the way to more understandable code.

Reading the above, you may be thinking I'm trying to convert you to FP, but that's not the case. I'm not one of those functional folks, and I wouldn't argue for a full-fledged FP. But at this point, hopefully, you're already thinking: "There should be something in functional programming. Is there any chance we can adopt any of its practices to simplify our own code?" I believe the answer to that question is, "Yes, we should strongly prefer immutability over mutability." It's mostly the lack of side effects and *immutability* that makes functional code easier to reason about. While many of the mainstream programming practices were originally developed in the context of FP languages,[85] unfortunately, arguably the greatest idea behind FP is still not quite popular. Immutability is what I truly discovered for myself only after joining Google, and I still wonder why people talk about it so little. It shouldn't be some sort of secret wisdom. So, let's take a closer look at what "strongly

[84] Yes, they're often not that product focused, but the only thing we care about in this context is their passion for understanding, because ultimately, it has a potential to serve the product (see Figure 1.1).

[85] For example, type inference, lazily evaluated streams, and generics [Seibel09].

preferring immutability" means in practice by examining the three pillars of what I would call semi-functional programming.

6.1. Strongly prefer immutable objects

Looking at an object within a function, especially within a longer one, it's not always obvious where and how it changes. Is it being modified in the lines above? Or below? If there are multiple changes, are there any dependencies between them? Does the order of these changes matter? Finally, if this object is being passed as a parameter to another function, can it be modified there as well? That's a lot of dependencies and even obscurity to deal with. And that's where immutability comes to help. As a rule of thumb, have as many objects in your code immutable as possible. Don't be too rigid about it, as there are definitely cases where mutable objects are handy, but it's much easier to reason about a piece of code if the vast majority of the objects involved are clearly immutable.[86] As Joshua Bloch points out in *Effective Java*, "Classes should be immutable unless there's a very good reason to make them mutable. [...] If a class cannot be made immutable, limit its mutability as much as possible" [Bloch08].

It's worth noting that you want immutability to be enforced. Given that an object's immutability is a part of its contract, it should be *internally verified* (see Chapter 3: "Design by Contract"). Most modern languages support that—at the very least, there are libraries providing immutable collections and simple ways of creating custom immutable objects (e.g., records, Guava,[87] and Immutables[88] for Java[89]). You want things to break loudly if there is an attempt to modify an immutable object—better at compile time or, at the very least, at runtime.

---

[86] Not to even mention the multithreaded environments, where immutable objects are priceless due to their inherent thread safety.
[87] `https://github.com/google/guava/`
[88] `https://immutables.github.io/`
[89] Project Valhalla would make it even better for Java CPU-bound applications. `https://openjdk.org/projects/valhalla/`

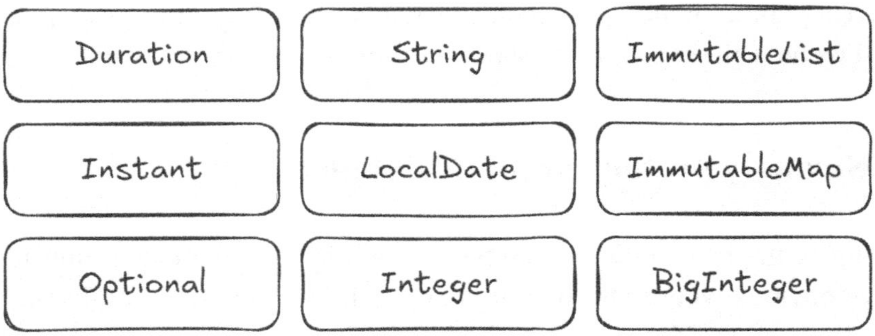

Figure 6.1. Examples of immutable objects in Java.

Immutable objects and immutable references are different

When talking about immutable objects, we're not talking about immutable fields or local variables (e.g. `const` in JS or `val` in Kotlin). Immutable variables are nice, especially in longer functions, but keep in mind that they usually don't make the objects themselves immutable.

Similar to objects (and for the very same reasons), prefer collections of objects to be immutable as well. However, once you switch to immutable objects and collections, you'll quickly realize that although they make the data flow clearer, manipulating them may result in a lot of boilerplate code. That's where transforming programming comes to help, which is basically using streams with `map`/`flatMap` operations:

```
var userDraftsToSave =
    users.stream()
        .map(u -> u.toDraft())
        .peek(u -> u.setTeamName(newTeamName))
        .toList();
```

In contrast, here is what the same would look like without streams:

```
// Naming it userDraftsToSaveBuilder would make this
// code block even longer:
var userDraftsToSave = new ArrayList<UserDraft>();
for (ImmutableUser user : users) {
  var userDraft = user.toDraft();
  userDraft.setTeamName(newTeamName);
  userDraftsToSave.add(userDraft);
}
```

The main downside of using streams is that to read and modify code that mixes classical imperative code blocks with declarative transformations, you need to be comfortable with both styles, which certainly requires a mindshift. As *The Pragmatic Programmer* puts it, "Thinking of code as a series of (nested) transformations can be a liberating approach to programming. It takes a while to get used to, but once you've developed the habit you'll find your code becomes cleaner, your functions shorter, and your designs flatter" [HT19].

6.2. Treat function parameters as read-only

Functional programmers love to talk about pure functions, referring to their resemblance to math functions. In mathematics, functions simply take arguments and return the "computed" result. They naturally don't modify anything, including their arguments. And it's a similar story with pure functions in FP—they're free of side effects by definition. Although having truly pure functions in the world of software engineering is often an overkill, it's a very helpful convention not to modify function parameters. Assuming that most of the parameters are immutable anyway, in practice it means just treating the rest of them as read-only.[90] If you follow this convention, then in a method call, changes are limited to the object's state (and transitively to the state of other objects that are part of the object state; see Figure 6.2).

---

[90] I know that it may look extremely suspicious to rely on a convention in this case, but surprisingly, this kind of "immutability" can be reliably achieved even without any assistance from compiler or assertions. See the "Enforced rule or a convention?" block below.

```
class A {
  // ...

  // Mutable objects:
  private B b;
  private C c;

  void foo(var param1, var param2) {
    // The following call potentially can modify the state of b.e:
    b.bar(param1, param2);
    // The following call potentially can modify the state of c:
    c.baz(param2, ...);
    // ...
  }
}
```

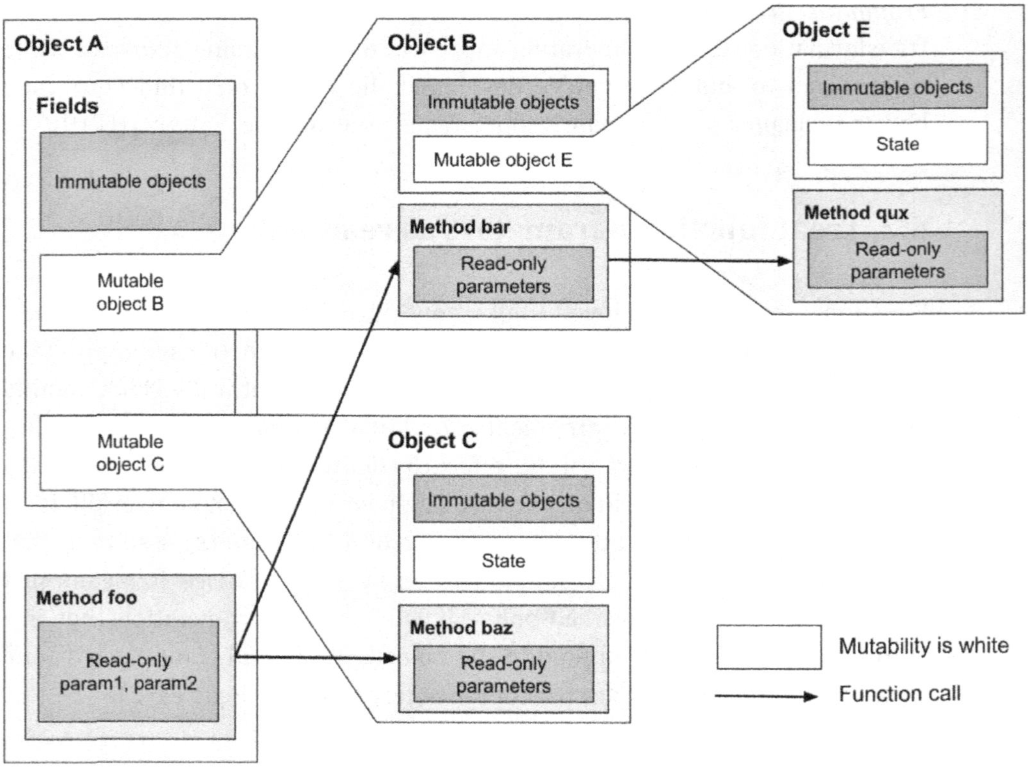

Figure 6.2. Objects potentially modified (in white) as a result of a `foo` method call, assuming all the parameters are read-only.

Looking at Figure 6.2, you may wonder whether limiting mutability actually makes much difference, given that there are still all those objects with a mutable state there. The point here is that if there is three times less mutability, that may be an order-of-magnitude cognitive load gain during the debug or code review. And treating function parameters as read-only plays an important role in that mutability reduction.

Pause and Think

For me, it's hard to imagine a parameter being modified inside of the function body, unless its name has a "Builder" suffix or any other clear indicator that it may be modified within a function call. As with most other rules, there are some exceptions to "you should treat all the parameters as read-only," but those exceptions should be very explicit.

Enforced rule or a convention?

Generally, enforced rules are far better than conventions, because enforcement makes it harder, if not impossible, to break a rule in production. In contrast, having a bug just because a line of code violated a coding convention is way too common. For example, let's consider the "parameters are non null by default" convention. If not enforced, it's super easy to end up with null being passed as a parameter value. Yet, it's primarily because this convention has to be followed on the caller side, so every new function usage has a potential to break it.

However, although non-enforced conventions are commonly fragile, it's a whole different story with the "treat parameters as effectively immutable" rule. This convention is kind of encapsulated inside of a function body, and normally, it's crystal clear whether you're modifying a parameter inside of a function or not. I've been programming within this paradigm with no enforcement for many years now, and I don't think I've ever seen a bug like "function accidentally changed a parameter that wasn't explicitly mutable." You don't have to be super disciplined to avoid such obvious mistakes. But of course, it's much better to have all the parameter objects truly immutable.

6.3. Make all the mutability explicit

At times, modifying objects may be handy or even unavoidable. You can even modify an immutable object—just explicitly transform it into a mutable builder, make the necessary changes, and then convert it back:

```
// User object is immutable:
var userDraft = user.toDraft();
userDraft.setName(newName);
ImmutableUser updatedUser = userDao.update(userDraft);
```

However, it's important to make such mutability explicit to the reader. As already mentioned in the previous section, explicitly labeling mutable function parameters is especially important, as function calls are more obscure by their nature:

```
public class Service {

  public void configure(Map<String, String> parametersBuilder) {
    // Because of its name you expect parametersBuilder to be
    // modified within a function call.
```

It's much easier to reason about a piece of code if the vast majority of the objects involved cannot change, and those that can stand out. That's the world I've been living in for the last ten years, and I don't want to go back into the wild west of "every bit of data may change at any given point of time."

> It's much easier to reason about a piece of code if the vast majority of the objects involved cannot change, and those that can stand out.

Summary

The number of moving parts is often a huge cognitive load issue by itself. But what makes the moving parts problem even more serious is obscurity around which potentially moving parts are actually moving, and in which scenarios. That's why sticking to the immutability principles of semi-functional programming greatly simplifies reasoning about code. In larger systems, you can even say it makes reliable reasoning possible. Here are the three principles of semi-functional programming:

1. Strongly prefer immutable objects and collections.
2. Treat function parameters as read-only.
3. Make all the mutability explicit.

Semi-functional programming is all about immutability by default. This is certainly so for objects and collections, and maybe even for the variables themselves as well. It's about functions that don't modify their parameters. However, at the same time, it's about being flexible enough to allow mutability where it's handy, but ensuring such mutability is explicit.

Those who love functional programming may argue that it's not functional enough to be called semi-functional. And those who strongly prefer the flexibility of imperative style may say that this approach is too restrictive and *too* functional. But that's okay—it's on the spectrum somewhere, in the spirit of pragmatic excellence.

Chapter 7

NAMING: THE OBSCURITY CHAMPION

You've probably already heard this old programming joke, but I can't stop myself from using it to open a chapter on naming: "There are two hard problems in computer science: cache invalidation, naming things, and off-by-one errors."[91] Although I'm not that sure about off-by-one errors, naming things most definitely deserves its place on the list!

Finding the right name can be truly hard, but sometimes, it's hard only because it is illusively simple. Let me start with a confession. This book grew out of my lectures at ITMO University, and as a part of the negotiation process to start that course, I sent a calendar invite to an associate professor titled "Chat about ITMO lectures." Guess what? While that title made total sense to me, for him, half of the meetings in his calendar could have been named the same! I'm still amused by the fact that he accepted my invite. That was a terrible mistake to make for someone applying to teach code simplicity, as good names are such an important component of code health.

It was the same year that I was sitting in a room with ten other people, many of whom I didn't even know. I believe we were trying to simplify some sentence structure when I mentioned that "I work with words a lot on a daily basis." Then somebody, who turned out to be a professional linguist, got intrigued and asked me about my occupation. Up to that point, she was obviously assuming she was the only expert in this field among us. Actually, I was surprised as well—me articulating it felt like a revelation to myself, too. But it's so true!

[91] https://martinfowler.com/bliki/TwoHardThings.html

Software engineers should be sort of professional linguists. We should always be on a journey to learn how to name things better and how to make our code read more like a story.

7.1. Name different things differently

It may sound like an obvious question, but should you ever use the same word to mean two different things? In conversations, it's absolutely unavoidable as some words have literally hundreds of different meanings. And that's cool because you can play with words to make your conversations more fun and your ads more sticky. Yet most of the time, we'd prefer others to easily understand what we're saying, and avoid any ambiguity, right? Thankfully, context comes to the rescue, as it typically clarifies what the word actually means within that specific context. And it's a very similar story with code, the difference being that in code, we have a really high bar for clarity. In fact, that's the golden rule of naming in programming: a name in its context should clearly communicate what that thing is or what it does.

> That's the golden rule of naming in programming: a name in its context should clearly communicate what that thing is or what it does.

Avoid using overly generic words

Generic words such as "Data," "Temp," or "Util" are bad name candidates as they may refer to a huge variety of different things, and different things should be named differently. "Class" is also typically not the best word to use when naming things in code, even if you're working on a software for schools. The word "class" is likely reserved to mean something else globally in your codebase, and again, different things should be named differently. In this particular case, even your compiler might suggest that!

Avoid comparisons to previous versions

Don't add adjectives such as "New," "Good," "Fast," etc. Those only mean anything in comparison to something else (the old version of that functionality). In the world where we actually get rid of `Foo` and are left with `GoodFastFoo` but are looking at `GoodFastFoo2`, something has gone wrong.

But what should you do if the word you want to use to name a function, variable, or a class is already taken? What if it already means something else within the same context? There are several options then. You could try using a *synonym* instead but that may be confusing, especially for new team members. Or you could use a phrase—in a phrase, you can be more explicit regarding what that thing is or what it does. Or, finally, you could just add a new term to your glossary.

Pause and Think

The last option is discussed later in Section 7.4, so for now, let's focus on phrases and synonyms. Say we have a variable named `millisSinceStart`. Is it clear enough? Well, it depends. If it's declared inside of a class named `FootballMatch`, it likely means "the number of milliseconds since the start of that match." However, if that's a variable in a method that does some heavyweight computations, the immediate context would suggest it's "milliseconds since the computation started." Such ambiguity in code is never a good thing, so we better name it `millisSinceMatchStart` or `millisSinceComputationStart`, respectively, depending on what it actually means.

It would be especially confusing if both of these variables are in scope but they're named synonymously, for example:

- `millisSinceStart` and `millisAfterStart`
- Or `millisSinceStart` and `sinceStartMillis`
- Or if the author couldn't get creative, even `millisSinceStart0` and `millisSinceStart1`

It would be funny if it wasn't sad instead. Let me share a real-life story about one of the AWS[92] APIs. We were implementing an exponential backoff for a message queue (AWS SQS[93]) to retry on a message handler failure. To do that, we had to know the number of previous attempts to process a message. Luckily, there was a standard AWS SQS message attribute named `ApproximateReceiveCount` (no, interestingly problems did not come from the word "approximate"). So, we implemented it, wrote a bunch of tests, and deployed it, but to our surprise, there was no exponential backoff in practice! As it turned out, `message.attributes.get("ApproximateReceiveCount")` was always returning `null`. What we found was that `message.attributes` wasn't exactly the AWS SQS message attributes—it was our own `Message` class, and in its constructor, we were filtering out all the attributes with numeric values! Why would you name it `message.attributes` if that's just a *subset* of its attributes? There was even no comment on that field. But most interestingly, even after we removed that filtering, the `ApproximateReceiveCount` attribute still wasn't there! We were telling AWS to populate all the message attributes:

```
var request =
    new ReceiveMessageRequest().withMessageAttributeNames("All")
```

We checked the wire, and the `MessageAttributes` field was there, but there was no `ApproximateReceiveCount` attribute among them:

92 https://aws.amazon.com/

93 https://aws.amazon.com/sqs/

```
{
    "Messages": [
        {
            // ...
            "MessageAttributes": {
                "AttributeX": {
                    // ...
                },
                "AttributeY": {
                    // ...
                }
                // No "ApproximateReceiveCount"
            }
        }
    ]
}
```

Guess what? The AWS SQS message has two different fields: `attributes` and `messageAttributes`. Some of its attributes reside in `message.messageAttributes` and some in `message.attributes`. From our perspective, it was even worse than `message.attributes0` and `message.attributes1`! When you see `message.attributes`, how on earth would you know there is another `message.messageAttributes` field that you should ask for to be populated in AWS response?

Finally, we configured the AWS SQS request to populate both `message.attributes` and `message.messageAttributes`:

```
var request =
    new ReceiveMessageRequest()
        // ...
        .withMessageAttributeNames("All")
        .withAttributeNames("All");
```

Only then we started receiving both fields in the AWS response and could choose in which of those particular fields the attribute of our interest resides.

There should be no such obscurity in code.[94] If there is a semantic difference between the two things, they should be named truly differently.

7.2. Make it clear and concise

While clarity is the most important metric you should optimize names for, there is another one that is almost as important: names should be concise. Obviously, there is a tension between these two objectives as often the name's clarity increases as its length grows.[95] Don't feel obliged to use short names only! Such typically self-imposed restrictions may be harmful. While it's good to keep a name as short as possible, you should feel free to use a longer one if that helps avoid confusion. Don't sacrifice clarity by overly optimizing for brevity.

> While brevity is important, clarity is the primary metric to optimize names for.

Let me share a story from my uni days. We were doing a thesis defense dry run with a bunch of other undergrad students, and when I finished presenting, our scientific advisor asked me, "Well, that was nice, but don't you think that it may be too simple for a thesis?" Fortunately, the very same year, I took a philosophy of science class, and one quote[96] instantly popped into my mind. As far as I remember, I replied, "Contrary to a common belief, it's not a complex thought that is scientific, but a clear thought." After several seconds of silence in the auditorium, my scientific advisor turned to other students and told them, "You

---

[94] Regardless of the historical reasons and backward compatibility concerns, the (chronologically) second field must have been named differently here.
[95] That's not always the case—we just saw how AWS folks used a longer `message.messageAttributes` name that didn't clarify anything. But having an option of using a longer name definitely opens the doors for a clearer name with the right level of detail.
[96] Now, I couldn't find the original author, but I'm 100% sure I didn't invent it.

should all memorize this quote as you may need it during your thesis defense as well." There were no more questions. And if clarity is such a virtue in academia, it's even more so in software engineering. Looking at a name in your code, ask yourself, what image does it create in the reader's mind? Does it have the right amount of details for the best readability? Or is there some other name that would paint a clearer picture? Never forget that while brevity is important, clarity is the primary metric to optimize names for.

7.3. Leverage the power of context

The key to making names both clear and concise is to leverage the power of context. Basically, there is no need in naming a field `userTeams` within a `User` class:

```
public class User {

  private final ImmutableSet<Team> userTeams;

  public void validateUserTeamNames(
      ValidationMode validationMode) {
    var userTeamLowerCaseNames =
        userTeams.stream()
            .map(team -> team.name().toLowerCase())
            .toList();
    // ...
  }

  // ...

  class UserBuilder {
    // ...
  }
}
```

Pause and Think

And yes, sometimes even notorious one-letter variables are clear enough within their tiny scope. While in general you shouldn't name a function `calculate()` that should have been named `calculateMonthlyInterestRate()`, it's a totally different story if it's a method of a `MonthlyInterestRateCalculator` class: `monthlyInterestRateCalculator.calculate()` looks great!

Obviously, to be able to leverage the power of context you need to have a meaningful context in the first place. That's where a rich domain model comes to help (see Chapter 5: "Independence Quest"). It provides you with a context to use, opening the doors for names that are both clear and concise.

7.4. Build your glossary consciously

Let's say a function has a long name—a name that sounds more like a phrase or even a sentence. You try thinking of a shorter alternative that would be clear enough, but you can't. Yes, using a longer name is absolutely an option, but if the same sequence of words appears in your code again and again, it may suggest that there is a new term to be added to your project's glossary.

As an extreme example, if you start seeing a repeating pattern of `nameAndSurname` in your codebase, you likely need to start using `fullName` instead. Full name is hardly even a term, as everybody knows what it means, which is fantastic but rarely possible. A more realistic example would be introducing the term "ACL." If you don't know what ACL means, that's absolutely okay! However, if you're working on authorization, having a term in your glossary that means "the list of users that have access to the item and the operations that they can perform on it" would simplify your life a lot.

Your project's glossary is a set of terms used in your codebase that a typical programmer would either not know or would expect them to mean something else.[97] Obviously, there is a limit on how big your project's glossary can reasonably be. So adding a new term comes with a cost as your glossary may grow out of control, not only complicating newbies onboarding but also causing high cognitive load for everyone in the team. A huge glossary is a commonly unfamiliar glossary. However, if you completely abstain from growing it, that would result in a dominance of long phrases—and not only in your code, but in your conversations as well, decreasing your code's readability and complicating your communications in general (see Figure 7.1).

> A huge glossary is a commonly unfamiliar glossary.

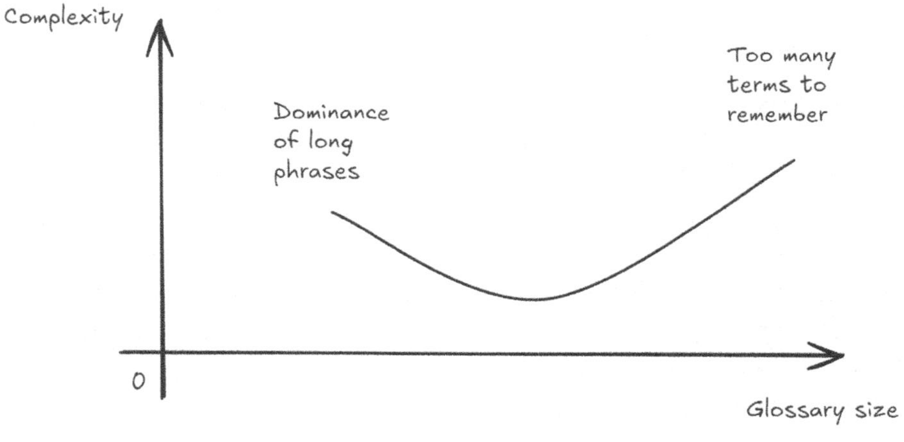

Figure 7.1. The optimal glossary size.

[97] By glossary here, I don't necessarily mean a formal document with a list of terms and their definitions, although that may be a good idea.

Actually, that's another benefit of introducing a new term in code. That term may easily cross the codebase boundaries, get into your documentation and conversations, and help you understand your domain better. As Eric Evans puts it in his classical *Domain-Driven Design* book, shared (or ubiquitous, as he calls it) language enables the "interplay of knowledge and ideas that lead to deep model insights" [Evans03]. Interestingly, this is very similar to any scientific progress as it relies heavily on the use of scientific terms. It may look mysterious to outsiders when they hear something like: "Flow of charge through a circuit can be measured by current and voltage," but it's this highly optimized language that enables insights scientists wouldn't get otherwise. So don't be afraid of adding new terms to your glossary as necessary. Typically, by doing so, you're just naming terms that conceptually already exist in your communications anyway. By giving them names, you're making your communications more concise and less ambiguous.

Pause and Think

Finally, terms shouldn't have a permanent residence in your glossary. Adding a new term is not an irreversible decision, as it may be there for a while, do its job, and sink into oblivion. Same as with software in general, your project's glossary is not a static thing—it evolves together with your software. It will both grow and shrink as some terms organically become obsolete over time. If you feel like it's time to abandon a term, abandon it. If you feel like you need to rename it, rename it.[98] Keep refining your glossary as your understanding of the problem domain evolves. It won't happen automatically, and it's not even easy to do, but it may be better to invest some time in improving your glossary than struggle with confusing names forever.

---

[98] That obviously requires a conscious effort, including changes in your codebase. For example, if you want to stop using the word "brand" in your conversations, deleting or renaming class `Brand` is absolutely a must. We all would love it if that happened automatically (and maybe in the future AI will mostly automate it), but unfortunately, all we can get from the code is the smell of an obsolete concept that somehow became more of a burden over time.

7.5. Use jargon

While naming things sometimes involves inventing new terms, in most cases, you can and should avoid doing that. Of course, default to words and phrases that everyone can understand, but when their expressiveness is not enough, it is *jargon* that comes to help—words that are widely known in a scientific, business, or tech community, or even those terms widely known in a community around a particular programming language, framework, library, etc. (see Figure 7.2).

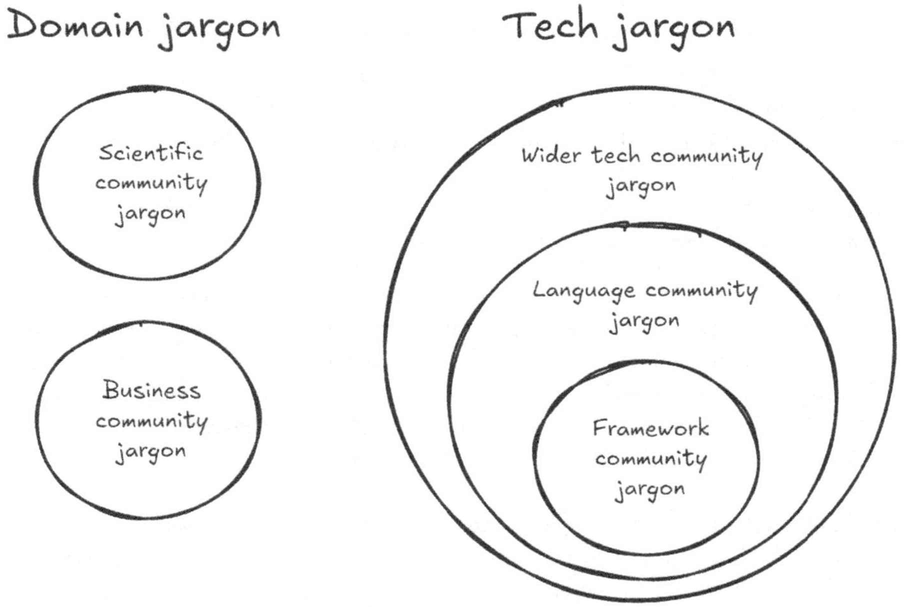

Figure 7.2. Domain and tech jargon.

Java-specific jargon example

Static import in Java is a way of referring to static methods without specifying the enclosing class. For example, `Preconditions.checkArgument` from Google's Guava library[99] is a widely known static method within the Java community that validates function parameters and is likely to be used in every second function in your codebase (if you follow the advice from Chapter 3: "Design by Contract"). If that's the case, there is not much reason to write `Preconditions.checkArgument` all the time—instead use static import and omit the redundant `Preconditions` prefix when making `Preconditions.checkArgument(...)` calls. Just say `checkArgument(...)`.

Yet by no means do I say that the whole engineering glossary (as reflected in your codebase) should be understandable by managers or domain experts. Some terms from your glossary are simply implementation details. For those, you may benefit from technical jargon—ideally as widely accepted as possible, but even widely known just within your company should be fine, too.

Naming—A collaboration opportunity

Naming things is a superb opportunity for collaboration with absolutely everyone, as it's less about knowledge and experience, and more about a fresh perspective. It's one of the touch points with less technical folks—brainstorming together will make them feel included, and you'll see how enthusiastic they'll get! It would also help you develop a ubiquitous language for your project—a common language shared by engineers, all sorts of managers, analysts, and business people [Evans03].

Also, if you're doing code review and notice a questionable name, suggest an alternative—it's one of the least offensive comments you can leave there! Even if you cannot think of a specific alternative, just mention that it's not clear enough—brainstorming names together is a fantastic collaboration opportunity!

---

[99] `https://github.com/google/guava`

7.6. Take it seriously when it is serious

Don't take all the naming indiscriminately seriously. Similar to interfaces, there are names that matter much more than others. Some concepts are core to your business and are ubiquitous in your codebase, while some are encapsulated within a single package, class, or even a function. Specifically for names in interfaces, the same rule of thumb from Chapter 4 applies:

```
private method
  < public method
    < public method in a public class
      < company-wide API
        < public API
```

Names in public APIs not only likely affect more people but are also de facto immutable,[100] so they should be as clear as possible. Overall, be pragmatic and use your time wisely: abstain from heated naming discussions on some local variables, but don't give up (easily) on stuff that matters and has lasting effects.

7.7. Listen to bad names

A good name is just a final touch. When you see a name that is not clear enough, it's not necessarily that somebody did a bad job at naming—it may well be a smell of a problem with the software design. In fact, it's one of the most reliable signals that this variable/function/class/service shouldn't exist in the first place. Unless you get the decomposition into modules right, it's extremely hard to avoid giving confusing names. It may well be the case that naming is "one of the hardest things in computer science,"[101] primarily because of design issues (see Section 5.3).

---

[100] Say hi to `message.messageAttributes` above.

[101] `https://martinfowler.com/bliki/TwoHardThings.html`

> When you see a name that is not clear enough, it's not necessarily that somebody did a bad job at naming—it may well be a smell of a problem with the software design.

7.8. Name same things the same

"Oh, you mean X!"—I'm sure this feeling of relief mixed with a mild frustration is very familiar to everyone and especially to those who have switched teams recently. Synonyms are a very common source of miscommunication. You know something as Y but the code says X. Well, sometimes it says Y, and sometimes X. For example, in our codebase, the same operation of linking an OAuth account was referred to as `linkOauthAccount`, `addOauthAccount`, `addOauthAccountForUser`, and `connectUserOauthAccount`. It's partially because, historically, we didn't support sharing OAuth accounts between users; however, even without such historical reasons, when you're naming something, you may not even be aware of an existing name or simply forget. However, once you've discovered the synonyms, it's important to find time to consolidate them. Inconsistent naming is confusing—the same things should be named the same.

Summary

Starting our conversation on naming with "Name different things differently" and finishing it with "Name same things the same" is symbolic. There is no rocket science in naming. But it is illusively simple. That may be the reason why I've never seen a self-development goal like "This quarter, I want to master naming things well." It just doesn't sound ambitious enough! Instead, there would be goals such as mastering a new framework, production monitoring, integration tests, infrastructure setup, and so on. However, in practice, having a not-that-measurable goal of mastering complexity management (and specifically mastering naming) would have a bigger impact on your overall productivity and your career as a Code Health Guardian. The good news is that you don't need to spend a lot of time expanding your vocabulary to be better at naming. Just being more conscious of and collaborative about it makes the biggest difference.

Part II

The Key Code-Related Processes

Chapter 8

TRADE-OFF ANALYSIS AND DESIGN DOCS

This book tells the story of a healthy codebase. So far, we've talked about the code itself—complexity problems, their causes, and various code simplification techniques. However, it would be only half the story if we stop here, as code health starts outside of code. Even the cleanest code possible can still be confusing if the technologies used, the underlying approach, or the overall architecture are supposed to be fundamentally different.[102] That's why the rest of the book covers the *key code-related processes*: trade-off analysis, design docs, build-versus-buy decisions, and code review. Without them it's hardly ever possible for a nontrivial codebase to be truly healthy.

> Code health starts outside of code.

Although up-front design is no longer mainstream,[103] over the last decade, I've been fortunate to work for several companies that not only use design docs, but also view them as one of the key artifacts of the development process. The main thesis of this chapter is that they are right—for most projects, design docs are objectively the best way of doing trade-off analysis. Design docs are mostly about effective collaboration and better-informed decision-making, but their

[102] For example, just recently, I wrote a piece of super confusing code in an attempt to get some data consistency guarantees without a huge performance hit out of the database that doesn't provide it out-of-the-box.
[103] The unpopularity of design docs today makes me even more eager to bring up this topic.

benefits don't stop there. In this chapter, we'll explore a wide range of their benefits and compare design docs with the alternative approaches.

Even if you're already familiar with design docs and practice writing them, I wouldn't recommend skipping this chapter—at the very least, skim through the headings and read the Summary section. I personally find it extremely helpful and motivating to remind myself of a full spectrum of benefits that design docs, if done properly, offer. Few of us are truly immune to the temptation to skip the proper trade-off analysis step and jump straight into the coding.[104,105] Reinforcing the best practices may be as valuable as learning them in the first place.

8.1. What are design docs?

There was a time when you could hardly avoid designing software. Developers were habitually waiting for days to run the next version of their program. So, they had really strong incentives to think up-front about their design, its trade-offs, and possible alternatives. Add the previous dominance of the waterfall development model, and you'll see why up-front design wasn't seriously[106] questioned for the most of the 20th century.

Yet, this changed in the 1990s with the rise of Agile. By that time, developers were no longer waiting for their turn to run a program on a mainframe and, even more importantly, the significant benefits of delivering software in short "build and feedback" iterations became widely recognized. As the Agile manifesto later summarized this mindshift: "We have come to value

---

[104] And this is mostly because it's so easy to believe that our latest idea is definitely the best one.
[105] I'm not referring to true prototyping here. A true prototype is 100% complimentary to design docs [HT19, "Prototypes and Post-it notes"].
[106] Even though there were individuals and even movements such as evolutionary design, which tried to challenge the "Big Up-Front Design," these ideas didn't get much traction at the time.

responding to change over following a plan."[107] Indeed, that may sound like an existential threat to design docs. If following a plan is deprioritized, is it still worth having a thought-through plan? Let alone having a plan in writing! In response to that, Fowler published his famous "Is Design Dead?" article,[108] where he argues that Agile doesn't kill software design but simply changes its nature. However, I don't think enough of us have truly embraced that view in practice since. Even though the extreme notions of design driven by tests and code only are long gone, it looks like most teams today still don't practice writing design docs.[109] The mere fact that so many developers confuse them with documentation speaks volumes.

Although some companies may use a stricter definition, I find the following loose definition of a design doc to be both liberating and empowering. *Design doc* is a collaborative document with the project goals, proposed solution, and trade-off analysis, where discussions happen and final decisions are made. Trade-off analysis (also known as pros & cons analysis) is at the very heart of this definition. The project can be a small new feature that takes a couple of weeks to implement, or something much bigger, such as a new microservice that takes months. The level of detail and length of the respective design docs will be different, but by definition, both of them should include some form of a trade-off analysis—at the very least for the proposed solution itself, but normally for the alternatives considered as well.

---

[107] `https://agilemanifesto.org/iso/en/manifesto.html`

[108] `https://www.martinfowler.com/articles/designDead.html`

[109] However, given that history is cyclical, it wouldn't be surprising if design docs once again become mainstream.

Project X

Status: Draft
Author: artie@intoperf.com
Last Updated: 2021-02-11

Objective

Background

Goals

Non-Goals

Overview

Detailed Design

Caveats

Privacy

Security

Storage

Figure 8.1. Table of contents for a sample design doc.

To give you an idea of the possible design doc structure, Figure 8.1 shows a table of contents for a sample design doc. Also, to even better understand what design docs are, it may be worth saying what they are not: *design docs are not documentation*. Not only do they focus more on implementation details, but what makes them *radically* different is that unlike documentation, they shouldn't be kept up-to-date. Design docs should never be updated once they are in the final status.[110] They are more like commits—once they're done they are effectively read-only. For example, let's say you're implementing a new

---

[110] Strictly speaking, design docs shouldn't be supported after the *project* is completed, not after the doc is finalized. In practice, the final status may sometimes be assigned prematurely as there may be unforeseen issues encountered during the implementation phase that would lead to the design doc correction.

system to store and manage some kind of resources. If a year later you decide to migrate those resources to a new database, that change would surely invalidate parts of your original design doc, but that is absolutely fine! The only changes allowed in old design docs are explicit "Not true anymore" comments, ideally with a link to a new design (see Figure 8.2). However, such comments are by no means mandatory. While it's reasonable to expect documentation to be up-to-date, never expect design docs to perfectly match the current implementation. Again, design docs are completely different in this respect.

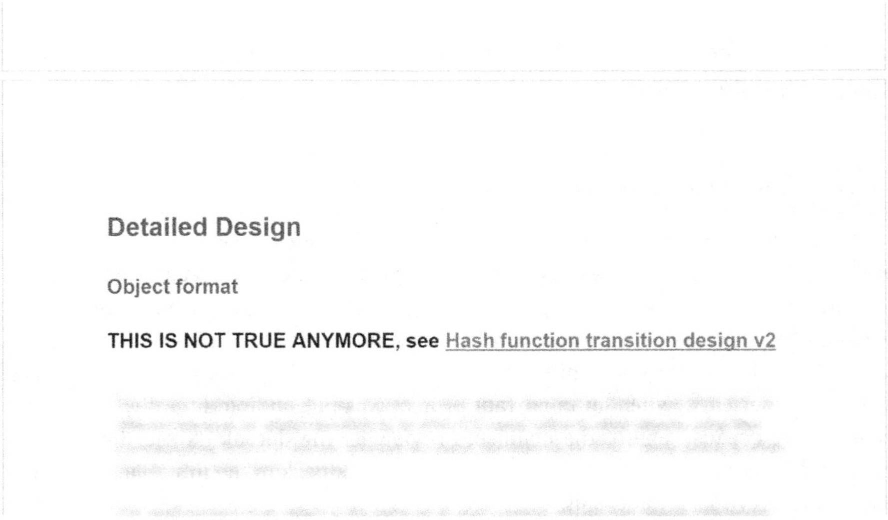
Detailed Design

Object format

THIS IS NOT TRUE ANYMORE, see Hash function transition design v2

Figure 8.2. "Not true anymore" comment example.

Finally, a couple of words about the process. Let's say you were given a task or project to lead. Although it may differ from one team to another, here is a typical workflow:

1. Collect the requirements from the key stakeholders.
2. Draft a design doc.
3. Collaborate in the doc with the stakeholders (aim at closing all the major comment threads).
4. Share the design doc with your team (same here: try to close all the major comment threads).
5. [Optional] Go through a formal design review meeting.
6. Change the design doc status to final and proceed with implementation.[111]

No rocket science here—just a few simple steps to unleash the power of collective intelligence.

Don't overfocus on reaching the consensus

I still remember a piece of advice I received when I got stuck with my second design doc at Google: "You won't make everyone happy here. Just get enough feedback and move on. Ultimately it is you who is responsible for this project." Getting as much alignment with others as possible early on definitely helps, but overfocusing on it may paralyze you, which is rarely a good thing. Interestingly, at another company, I had a very different experience, when an important project was put on hold literally for a half a year simply because one of the engineers reviewing the doc didn't see the value in making it happen. Aim at asking the right folks for feedback in the first place, and take any feedback with a grain of salt.

[111] If during the implementation phase you realize that there were some major misses in the design doc, you may revisit the design and go through another round of review afterwards.

> **Don't give others EDIT permissions unless they need them**
>
> Once, after joining a new company, I quickly realized that there is a common practice of sharing docs with EDIT permissions by default. Given that I was new, for almost a year, I had been trying to always do the same, until one day, one of my colleagues silently changed a single word in my doc, causing three problems during the review by senior leadership. The principle of least privilege very much applies to design docs—give others only permissions that they actually need to accomplish their task. If you're asking them for feedback, grant them permissions to leave comments in your doc but not to edit it. That's what we used to practice in my teams at Google, and it worked perfectly.

8.2. No architects at Google, "just" software engineers

The content of this chapter was first shared as a "Design docs culture" talk at a conference for software architects in 2021. When organizers asked me if I could share something at their conference, I stumbled for a moment. Any content specifically for architects? On the one hand, by that time, I'd already accumulated quite a lot of architectural experience—building infrastructure, optimizing latency in large distributed systems, designing data consistency models, and so on. But at the same time, I never thought of myself as an architect, and it took me some time to realize why. At Google, every software engineer writes design docs, and in them, you make tons of decisions—some clearly architectural, some not—but the boundary here is pretty blurry,[112] and we didn't even think about it in such terms. When I realized what a dramatic effect the culture of design docs had on Google, I was resolved—that's what I was going to talk about at that conference—the culture of design docs—culture that has basically wiped out the word "architect" from our lexicon. In fact, it's one of the key benefits of writing design docs: when they become an organic part of team culture, they obliterate the artificial boundary between "ordinary" software engineers and architects. Just like every engineer writes unit tests,

---

[112] My favorite definition of architecture is "stuff that's hard to change." Yet what counts as hard may be different for different people (and on different days), so in practice, it's not always clear whether something is a part of a solution architecture or not.

the same way, every engineer writes design docs. It's neither an exclusive privilege of architects, nor is it their exclusive burden. Common activities, language, and best practices ensure that there is no division into PowerPoint architects and hands-on engineers! Brilliant, right? It is the first benefit of writing design docs: they have a truly powerful team-building effect.

SREs at Google

Just like in other companies, in Google's early days, there were tensions between software engineers and system administrators.[113] Yet, given how critical `google.com` reliability is for Google, low prestige of system administrators' work was posing a huge threat, screaming for a radically creative solution. The main part of the solution was to introduce a new role, which was called Site Reliability Engineer (SRE). SREs when being hired would go through the same famously tough interview process as software engineers, and as part of their job they would build (code!) their own tools.[114] In addition, SREs would be paid 10% more than other software engineers, and there would be a voluntary rotation program giving developers a taste of what it's like to be an SRE at Google. Finally, software engineers could easily move to an SRE position and vice versa. Thus, the distinctions between these roles were deliberately softened.

This move was a remarkable success! At the end of the day, Google is an engineering company, and its employees feel proud to be software engineers. Similar to how Google softened the differences between software engineers and architects with the help of the design docs culture, with the introduction of an SRE role, it made every system administrator a software engineer. Having an off-site together and a bunch of kudos instead probably wouldn't have the same team-building effect (you know I'm being sarcastic).

---

113 There were no DevOps at the time.

114 Also, SRE is explicitly capped at 50% "automatable toil," leaving half of their time to do coding/engineering activities that are explicitly focused on reducing toil/improving scale going forward.

8.3. Trade-off analysis

Although the team-building effect of design docs is an important benefit, it's definitely not the main one. The main benefit is that design docs are objectively the best way of doing trade-off analysis for nontrivial projects. Writing a design doc not only makes the delivery faster, but if done properly, it also makes the process itself more productive and delightful. I know that's a strong statement, so let's prove it!

> Design docs are objectively the best way of doing trade-off analysis for nontrivial projects. Writing a design doc not only makes the delivery faster, but if done properly, it also makes the process itself more productive and delightful.

First, a proper timely trade-off analysis, regardless of whether it's in the form of a design doc or not, always reduces the risks and costs involved. I'm sure you're familiar with a classical diagram of the relative bug cost depending on the development stage the bug was discovered in (see Figure 8.3). Although untimely or a completely missing trade-off analysis doesn't necessarily lead to bugs straightaway, it often results in suboptimal decisions, complex code, late pivots, or constant complaints about the implemented solution.[115] So, even if it's not rolled back, the costs may be huge (very similar to Figure 8.3). And if it is, it's not just a matter of costs anymore—truly, there are few feelings worse than seeing your work being deprecated in a matter of months. So it's hard to overestimate the importance of a timely trade-off analysis.

[115] If it were a book for HR/managers, I would definitely finish with: "And what about turnover among your best people?"

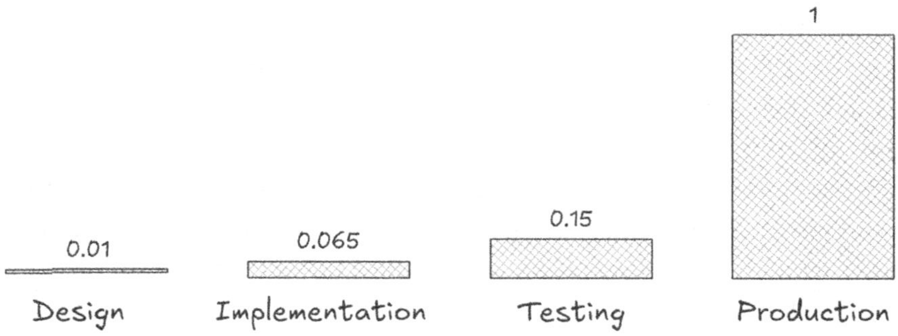

Figure 8.3. Relative cost of fixing defects.[116]

However, apart from design docs, there are several alternative trade-off analysis approaches, such as discussing a proposed solution with all the stakeholders (all together or 1:1) or simply thinking it through before jumping into coding. Even on-the-fly decision-making when you're already coding may be yet another form of trade-off analysis. However, what makes design docs such an attractive option is a combination of the following properties. Design docs are:

- Collaborative
- Centralized
- Standardized
- Explicit

The other forms of trade-off analysis also hold some of these properties, but it's design docs that so uniquely combine all of them:

Collaboration. Design docs promote both *timely* and *inclusive* collaboration. As part of the process, the author shares the draft with anyone who may be affected by it, including all the teammates—and that's before any code is

116 Numbers from IBM Systems Sciences Institute, `https://www.researchgate.net/figure/IBM-System-Science-Institute-Relative-Cost-of-Fixing-Defects_fig1_255965523`

written. First of all, that unleashes the power of collective intelligence applied at the right moment. But also, if somebody didn't speak up about their concerns at that point, they lose the moral right to complain about it later.

Centralization. Not having all the right people together in the same room is a common problem with meetings. This is especially true if you need a series of meetings with all the stakeholders. So in practice, with the meetings approach the discussion is typically fragmented, making it difficult for everyone to stay on the same page. It's not too uncommon to forget to inform Stakeholder A that you've decided to change one of the earlier decisions after discussing it with Stakeholder B. Design docs are collaborative online docs,[117] which mostly solves this problem as now there is an authoritative source of truth—it's the design doc.

Standardization. Design docs bring standards in several ways. They standardize timing, template, and a set of rubrics—topics that should be considered in a typical design (such as privacy, security, storage, and scalability). It may be a good idea to have a standard company-wide design doc template. A single[118] template makes the review process easier, as everybody gets used to the structure of the doc, fonts, and all those unimportant details that shouldn't add any cognitive load, distracting readers from the content. Having a list of rubrics in the template (for example, as headers; see Figure 8.1) is an opportunity to remind authors to think about important topics that are easy to forget about. Yet, beware of priority delusion. Have just *the most important* questions listed in the template, because if you have dozens of them, in practice, all of them will be ignored. Use the power of checklists,[119] but don't try to abuse it. As with most other lists, the list of design doc rubrics should be short to be useful.

Explicitness. The devil is always in the details, so writing things down and drawing diagrams typically helps a lot with clarifying our thoughts and

---

[117] Such as Google Docs or Canva Docs, for example.
[118] Or maybe two, just not a dozen of them.
[119] `https://www.newyorker.com/magazine/2007/12/10/the-checklist`

discovering flaws in our thinking. It may easily happen that the alternative you initially considered to be dumb after a diligent trade-off analysis ends up being the best way forward.

Now, let's formally compare design docs to their counterparts. In Table 8.4, there are different forms of trade-off analysis in rows and their desired properties in columns. Right now, I'd like to encourage you to pause reading for several minutes to fill in your estimates for each of the cells: to which extent do you think a particular approach holds a particular property?

	Timely	Familiar process	Second opinion	Inclusive	Everyone on the same page	Clarity	Level of detail	Documented	Direct cost, man-hours
"All-hands" meetings									
1:1 meetings									
In your head prior to coding									
When coding									
Design doc									
Short design doc									

Table 8.4. Exercise—Trade-off analysis of different ways of doing trade-off analysis.

In Table 8.5, you can find my take on the issue. I'm sure that if you've filled out your estimates in Table 8.4 thoughtfully, even though they surely differ from mine, you would still come to the same conclusion: there are no real competitors to design docs here. The second place is a distant second place, unless you heavily prioritize some of the desired properties over the others.

	Timely	Familiar process	Second opinion	Inclusive	Everyone on the same page	Clarity	Level of detail	Documented	Direct cost, man-hours
"All-hands" meetings	2	3	3	2	2	1	1	2	0
1:1 meetings	3	2	3	0	0	1	2	1	2
In your head prior to coding	3	3	0	0	0	1	2	1	3
When coding	1	3	0	0	0	2	3	1	3
Design doc	3	3	3	3	3	2	2	3	1
Short design doc	3	3	3	3	3	2	1	3	2

Note: White is good, dark gray is bad. In a book I have to use shades of gray, but normally you would use red/orange/yellow/green instead. You can find examples of the original color-coded tables at `https://codehealthguardian.com`.

Table 8.5. My take on the exercise.

Why color-coded tables?

I've first seen color-coded trade-off tables in internal Google documentation, comparing the pros and cons of different storage options: Spanner, BigTable, Firestore, and others. Their goal is to simplify trade-off analysis and the decision-making process by making it visual and avoiding duplication. The most popular alternative I've seen so far is a simple bullet list of pros and a separate list of cons for each of the options. But what has always confused me about such lists is the common DRY violations in them—the same argument would often appear (sometimes in different forms) in multiple lists. Thus, ever since I discovered color-coded trade-off tables, I strongly prefer them over the plain pros & cons lists. Making the summary of the trade-off analysis visual also simplifies both the decision-making process itself and communications around it. In this book, we've already used such color-coded trade-off tables in Chapter 4, and we'll use them in the next chapter as well.

When using trade-off tables in a real design doc, don't forget to write down a justification for every non-obvious estimate from the table. Also, if the columns significantly differ in priorities, you might want to explicitly communicate that and justify the priorities as well.

8.4. One-pagers

However, it would be a mistake interpreting the above results as an imperative to write a design doc for every project. To start with, there are trivial projects that don't require any substantial analysis. If there is a great straightforward solution, why waste time on a design doc? When there is nothing to analyze, strictly speaking you won't even be able to produce a design doc because, by definition, it must have some trade-off analysis as a part of it. You can still write a doc about your project, but it would be a stretch calling it a design doc.

Yet another scenario when writing a design doc may be unnecessary is when you've done something *very* similar in the past. In the previous case, there was no room for the trade-off analysis, and here, that analysis has already been done before. However, it's okay to skip the design doc step only if you're absolutely sure that what you've done before is *really* similar, and your context hasn't changed much since then.[120] Context changes may easily invalidate the results of an earlier analysis, making it necessary to write a new design doc.

Some would also argue that if you're an experienced software engineer, there is not much value in writing design docs—why not just have a design in your head instead? That's a much cheaper option than a design doc in terms of time and effort, and is almost as effective. I agree that senior engineers benefit from writing design docs much less than juniors, but the same is true for code reviews, and code changes from senior engineers are nevertheless being reviewed by somebody else, right? At the very least, you can view code reviews and design docs as safety nets. If you're a senior engineer, an explicit collaborative trade-off analysis may be rather useless in 9 cases out of 10, but that doesn't prove the safety net to be useless. Moreover, as you can see from the rest of this chapter, design docs are so much more than just a trade-off analysis tool, and the other benefits of writing design docs are still the same, regardless of your experience.

Interestingly, whether to write a design doc is not a binary decision. Do you remember that soul-quivering feeling from the past when your school teacher was asked by a fellow student, "And...how long should the essay be?" The good news is that for design docs, there is no minimal length. At least at both Google and Canva, short design docs[121] (commonly referred to as one-pagers)[122] are even often preferable. There is actually one more observation for us to make from Table 8.5: the best alternative to design docs is *short design docs*. As you can see, they provide almost the same set of benefits as normal-length design docs—we're just trading a lower level of detail for lower direct cost, preserving all the rest. That often makes writing a short design doc the best option. Specifically, in the cases mentioned above, if you lean toward not writing a design doc, consider writing a one-pager instead!

---

[120] Such as technology advancements or their availability, for example.

[121] No template, no rubrics.

[122] Yes, the name is a bit misleading because everybody understands that in fact we're talking about a two- to three-page limit really.

> If you lean toward not writing a design doc, consider writing a one-pager instead!

Finally, a design doc will never cover all the decisions you need to make as part of your project. Just because it will never have the same level of detail as the code itself. There are tons of smaller decisions you'll make only once you start coding, and probably even some of the major ones that happened to be in your blind spot when you were writing the design.[123] Don't be caught in the analysis paralysis[124] trap—once the level of detail in your design doc draft becomes good enough, just move on with the review and implementation steps!

8.5. Design docs, code, and tickets

So far, we've focused on the two main benefits of design docs—trade-off analysis and their team-building effect. Yet there are a couple more. First, design docs enable us to have diagrams in code comments. Yes, code provides many more details than design docs, but it doesn't necessarily make it better at explaining how the system works. You definitely saw those awkward-looking attempts to explain complex subjects in code comments. Yet, often, a single diagram has more explanatory power than a thousand words. To add a diagram to your code, just add it to your design doc and have a link to it in the code:

```
// That may be necessary for M(E)V/F cache optimization,
// for details see the design doc:
// https://docs.google.com/.../edit#heading=h.h4923hhdj4jvs
if (correctedStartInstant != null) {
```

---

[123] As already mentioned, for major ones, it's absolutely okay to revisit your doc during the implementation phase to add or update some trade-off analysis. You might need to go through another round of review after that, though.

[124] The state of overthinking a decision to the point where no action is taken.

Here, we rely on the fact that design docs, unlike documentation, are effectively read-only once they're in the final status (see Section 8.1), which makes them a perfect means for adding diagrams to code. Unlike wiki, you can be sure it won't give you any nasty "Not Found" surprises later on.

Furthermore, a design doc is a great way of explaining your project to others. You can think of it as a website—a good design doc has "Background," "Goals," and "Solution overview" sections, which together tell the story of your project. So, if somebody is interested in understanding what you're working on, sharing a design doc link is much more informative (and easier for you) than sharing a bunch of tickets. Design docs are a great means of raising awareness of your projects among colleagues.[125]

Finally, as a nice side effect, it may help you in the next performance review cycle, given that design docs clearly demonstrate your design, collaboration, and maybe even leadership skills. Hopefully, all the arguments laid out in this chapter are convincing enough to make you write a design doc for your next nontrivial project. If you've never done it before, start with a one-pager.

Summary

Design docs are the best way of doing trade-off analysis for most projects as they standardize the process and make it timely, explicit, collaborative, and centralized. In addition, they:

- Enable diagrams in code
- Have a team-building effect
- Serve as a website for a project
- Serve as an important artifact for the next performance review cycle

---

[125] If there are other teams involved, it may even help you align with them and get things done faster.

These reasons combined make starting any nontrivial project with a design doc a great default strategy. And whenever you lean toward not writing a design doc, consider writing a one-pager instead!

Although it's not strictly necessary for design docs to be a part of your team's culture to be beneficial to you, they shine when they are. So, if your team doesn't practice writing them, be a catalyst for positive change! Yet, keep in mind that design docs are just a tool to make you, your users, and your teammates more productive and happier. Not the other way around. It's easy to get overly excited and take it too far. That's why so many experienced folks repeat it twice that they mean one-pager when asking somebody to write a design doc.

Chapter 9

BUILD VS. BUY

If you're not new to software engineering, surely you know that only a few experiences match the satisfaction of deleting code, as less code in your codebase means less code to maintain. In terms of the formula of codebase complexity from Chapter 4 (see Eq. (1)), it's great when you can isolate complexity making the t factor small for a particular term, but removing that term entirely is even better. Sometimes, this can be achieved by eliminating unnecessary functionality; yet another option is to simply delegate that complexity and make it somebody else's problem.

This chapter is about reusing existing stuff: libraries, services, frameworks, components, and so on. It may be more relevant for senior folks as the high-stake build-versus-buy decisions often pop up to the top, becoming brain teasers for architects[126] and senior leadership. However, from time to time, all of us face the dilemma whether to implement something by ourselves or adopt an existing solution. In this chapter, we'll explore how to make better-informed build-versus-buy decisions.

[126] Yes, it's great if they disappear as a strata (see Chapter 8), but I'm using the word "architect" here as a synonym for "more senior folks," to not repeat the phrase that was already used in the previous sentence.

9.1. Your future needs: The pain of a buyer

Unfamiliarity and *unstable dependencies* complexity causes (see Figure 2.3) are often interrelated. Widely known frameworks, cloud computing platforms, components, and libraries are usually rather stable. It's virtually guaranteed they won't disappear tomorrow, and most of them guarantee backward-compatibility. Nevertheless, introducing a substantial dependency to our system, even on a widely known and stable module, is not necessarily a good idea. There is one huge caveat out there—we have to make sure it suits our needs, and not only today, but a year from now as well, which is the tricky part. Writing a proof of concept is typically enough to validate that our current needs are met, but what if several months later we realize that we need a feature that is not there? If that's a widely known module, likely any serious bug we encounter will be fixed reasonably fast,[127] but that's definitely not the case with feature requests. Even if they ever get implemented, that may be too late for us.

Yes, you can try to provision your future needs up front, but there is a nuance—humans are terrible at predicting the future. That's not to say that you shouldn't try. Take some time to project your future needs and talk to the potential stakeholders about it, but keep in mind that regardless of how hard you try, at the end of the day, it's likely you'll still be wrong.

9.2. Why reinvent the wheel

Another common strategy of addressing these risks is to buy premium support from a vendor, or if you consider depending on an internal module, to get explicit guarantees from the owning team.[128] Unfortunately, in practice, that also doesn't lower your risks substantially. As an extreme example, several years ago, there was an infrastructure service at Google with five to ten internal

---

[127] In most cases, you can easily test this assumption for any particular module or a vendor.

[128] Great relationships with that team may be the best kind of guarantee.

clients. Yet, our team was their only client with a launched product, and we had tens of millions of users at that point. I know that sounds absolutely weird (and it *is* weird), but one day, the team owning that service shut down some of the endpoints we relied on without even telling us. That story went all the way up to our CEO Sundar Pichai, and that team and their service were shut down in a matter of months. Yet for us, the damage from an ill-fated dependency was already there. Although that service wasn't widely known, it was five years old at that point and seemed to be mature and stable enough. As it turned out, it wasn't quite so.

A similar story happened to us on another team at Google. This time, we introduced a dependency on a photo storage service. Yet after about a year, Google Photos became so popular that our daily batch jobs executing on hundreds of machines couldn't finish processing their daily dumps in 24 hours. It was affecting the quality of our data, so we asked the Photos team for more granular dumps—a dump per tenant or something like this. Yet months went by, and they still didn't address our problem.[129] It just wasn't their priority. They were too successful and too busy.

We experienced very similar problems at another company as well. Our infrastructure was mostly based on Amazon Web Services,[130] which normally worked just fine, but every now and then, we were getting a huge latency spike for storage (EBS[131]) I/O operations, which translated into database issues. In turn, this caused service timeout and error spikes, typically lasting for less than 30 seconds. Although we had premium support with Amazon, whenever we tried to make them investigate the problem behind these sporadic EBS latency spikes, the answer was always the same: "Sorry, that's within the SLA."

There are too many horror stories of how dependencies on modules owned by others didn't play out well, illustrating how easily the risks we've discussed above transform into real problems. The good news is that there is an obvious

---

[129] As a quick fix, we just continued throwing hundreds of additional CPUs at it.

[130] `https://aws.amazon.com/`

[131] `https://aws.amazon.com/ebs/`

way of avoiding these risks—just build your own solution instead of depending on an existing one—an alternative better known as reinventing the wheel. It is a notoriously wasteful idea, but given that it abolishes so many risks by putting one in control, sometimes that may be the right thing to do.

How do you decide when to reinvent the wheel and when to adopt an existing solution? Later in this chapter, we'll sketch a proper trade-off analysis, but let's first acknowledge that many of us are biased toward reinventing the wheel. Searching for modules that would satisfy our needs and comparing their pros and cons may be tedious. On the contrary, as long as it's not too hard, reinventing the wheel is tempting because the wheel is by definition a well-isolated, well-defined piece of software, unlike much of our day-to-day work! It's also often accompanied by interesting engineering challenges. And finally, your wheel is something you can *truly own* and effectively learn from, growing as a professional. So, that's not a completely irrational bias. I would even say that it's a very reasonable bias to prefer reinventing the wheel over adding an external dependency. Yet, it's counterproductive. To better understand why, let's sketch a trade-off analysis.

9.3. Costs and risks overview

To make it more vivid and less abstract, it's much better to study the rest of the chapter when you're actually in the process of making a build-versus-buy decision yourself. Unless that's the case (or you're just curious), consider going to the Summary section and coming back to this chapter once you face a real build-versus-buy dilemma. If you choose to proceed anyway, don't say I didn't warn you. Take the following results with a grain of salt as it's a rather high-level analysis. Yes, we'll get some interesting and universally valuable insights, but most importantly, the below can serve as a starting point for your own build-versus-buy trade-off analysis.

Choosing between building your own solution or adopting an existing one may be hard. The main factors at play here are (represented as columns in Table 9.1):

- **Direct cost:** Implementation, buy, and integration costs
- **Migration cost:** The price to pay in case the adopted solution stops meeting your needs or becomes burdensome in the future
- **Future costs:** Usage and maintenance costs[132]
- **Future risks:** All the associated development, reliability, and security risks

<table>
<tr><th rowspan="2"></th><th rowspan="2">Direct cost</th><th rowspan="2">Migration cost</th><th colspan="2">Future costs</th><th colspan="2">Future risks</th></tr>
<tr><th>Usage costs</th><th>Maintenance costs</th><th>Develop -ment risks</th><th>Reliability and security risks</th></tr>
<tr><td>Build your own thing</td><td>High?</td><td>Not something you would provision for?</td><td rowspan="2">New interfaces to understand and remember</td><td>Significant?</td><td>Buggy?</td><td>?</td></tr>
<tr><td>Use an existing solution</td><td>Cheap?</td><td>?</td><td>Low or none</td><td>Hard to fix bugs and add new features</td><td>?</td></tr>
</table>

Note: White[133] means cheap, dark gray is super expensive. Question marks indicate no/rough estimates.

Table 9.1. Trade-offs of building your own solution versus adopting an existing one.

---

[132] Note that while the first two are one-off costs, these are not. In addition, usage costs, unlike everything else, typically affect more of your developers—sometimes even the whole fleet of them.

[133] In a book I have to use shades of gray, but normally you would use red/orange/yellow/green instead (and white for N/A and question marks). You can find examples of the original color-coded tables at `https://codehealthguardian.com`.

Table 9.1 is mostly about code-related costs, whether we're talking about developer onboarding; integration costs; or migration, development, or maintenance work. There are some non-code-related costs and risks involved as well, such as financial aspect, reliability, and security risks, but these are highly context dependent.

9.4. Build option: Costs and risks

Let's justify the estimates in the first row of Table 9.1 (if the estimates are more or less obvious to you, feel free to skip the following two sections as they're rather tedious).

Direct cost. Building your own thing may require a significant up-front investment. Sometimes you can start small and pay it in chunks over a long period of time, but changing a live system is often hard, so it usually increases the total direct cost significantly.

Migration cost. Migrating off your own solution is usually pretty hard as provisioning for it would have increased the direct cost even more. So in practice, that's rarely something you would provision for. You may be lucky, though.

Usage costs. While the solution itself would likely be developed by a small cohort of people, it often affects many. New interfaces for developers to understand and remember may be a significant continuous overhead (see Chapter 4: "Interface Simplicity").

Maintenance costs. These are usually significant. That includes providing all kinds of support to the solution adopters, maintaining documentation, maybe some infrastructure work, and/or even an on-call roster.

Development risks. More or less, everything is in your control when you reinvent the wheel, but that doesn't mean that development risks are necessarily low. If you don't have expertise in that particular domain, your implementation is likely to be buggy. Due to its limited adoption, many of those bugs may stay hidden for a long time, giving you an illusion that it's all good and making any future planning less accurate and the development process more disappointing.[134]

Reliability and security risks. Again, you're mostly in control here, but how good will it be in practice? We definitely tend to overestimate the security and reliability of our own solutions, in part due to their limited adoption. So, be ready for unpleasant surprises later on.

9.5. Buy option: Costs and risks

Now that we've talked about the first row, let's justify the estimates in the second row of Table 9.1.

Direct cost. Onboarding of an existing solution is normally relatively cheap. This is so especially if the module in question is widely known, as in addition to the integration costs, direct costs also encompass the costs of learning the best practices and developing your usage patterns. I know it sounds weird, but the money factor can also *contribute* to the overall cost. Yet, in practice, it rarely overturns the build-versus-buy decisions.[135]

Migration cost. That's 100% context specific. When the migration costs are extremely high, making the buy decision almost irreversible, it's called *vendor lock-in*.

---

[134] However, given that reinvented wheels are usually well-isolated and well-defined pieces of software with interesting engineering challenges, high programmers' engagement could compensate for these issues.

[135] There are exceptions of course, such as Google Spanner, which as of the time of writing is probably the best database out there, but not that many products can actually afford it.

Future costs is where adopting an existing solution option truly shines:

- ***Usage costs.*** Any solution brings new interfaces for developers to understand and *remember*, together with their usage patterns. Similar to direct costs, being a widely known module obviously lowers its usage costs as well.
- ***Maintenance costs.*** These are mainly integration costs and perhaps communication with the vendor support.

Future risks, in contrast, are the dark side of depending on existing modules:

- ***Development risks.*** Even though widely known modules are more likely to have all the features you'll ever want, if you're unlucky to want something that is not there, a module being widely known and stable just lowers your chances of getting your requests. The real-life stories from the beginning of the chapter are vivid illustrations of how dependency risks can easily transform into real problems.
- ***Reliability and security risks.*** This is highly context dependent. For example, if you're Google, it's likely that reliability and security risks of any external solution would be too high for you to accept. On the other side of the spectrum, for most companies, the security and reliability risks associated with using GCP[136] or AWS[137] are typically justified.

9.6. Why default to buy

If you focus on the future costs and risks only, it's easy to be misled that the build-versus-buy decision is all about the risk appetite, as the pros and cons of the two options seem to more or less mirror each other (see Table 9.2). Yet, the first two columns shift the balance quite a lot.

---

136 `https://cloud.google.com/`

137 `https://aws.amazon.com/`

	Direct cost	Migration cost	Future costs		Future risks	
			Usage costs	Maintenance costs	Development risks	Reliability and security risks
Build your own thing	...	...	New interfaces to understand and remember	Significant?	Buggy?	?
Use an existing solution	...	...		Low or none	Hard to fix bugs and add new features	?

Note: White means cheap, and dark gray is super expensive. Question marks indicate no/rough estimates. In a book I have to use shades of gray, but normally you would use red/orange/yellow/green instead. You can find examples of the original color-coded tables at `https://codehealthguardian.com`.

Table 9.2. Pros and cons analysis of building your own solution versus using an existing one: Focusing on the future costs and risks.

To give you some intuition why reinventing the wheel is a bad idea in most cases, let's demonstrate that if you choose to buy, then even the worst-case scenario down the road would typically be better than if you initially decide to build. So, let's say you decide to buy and half a year later realize that some critical features that you need are missing, or there is a bug you care about that's not being fixed for a long time. To make things worse, you can't find any alternatives on the market that meet your needs, so just several months after you made a decision to buy, you pivot and finally decide to build your own thing. The question is, should you regret your initial decision to buy? Surprisingly, not necessarily so! Security and reliability risks aside, let's compare the final costs to see how that is possible.

<table>
<tr><th rowspan="2"></th><th rowspan="2">Direct cost</th><th rowspan="2">Migration cost</th><th colspan="2">Future costs</th></tr>
<tr><th>Usage costs</th><th>Maintenance costs</th></tr>
<tr><td>Build your own thing</td><td>BUILD_DIRECT</td><td>-</td><td rowspan="2">USAGE</td><td>BUILD_MAINTENANCE</td></tr>
<tr><td>Use an existing solution</td><td>BUY_DIRECT</td><td>BUY_MIGRATE_OFF</td><td>~0</td></tr>
</table>

Table 9.3. Variable names as used in the formulas.

The cost variables below have the names as shown in Table 9.3. The only difference is that there are two build your own thing direct costs:

- **BUILD_DIRECT_ACTUAL:** The actual scenario, which is to start with a buy option and build it half a year later.
- **BUILD_DIRECT_ALTERNATIVE:** The alternative scenario, which is to start with a build option.

Note that it's perfectly fine to skip the math part below and go straight to the final paragraph of this section. And by no means do I expect any mathematical calculations to be involved in a typical build-versus-buy decision-making process. In fact, I would be quite surprised if you could ever accurately quantify the costs and risks involved beforehand.

The actual costs are[138]

```
ACTUAL_COSTS = BUY_DIRECT + USAGE * n_months
+ BUY_MIGRATE_OFF + BUILD_DIRECT_ACTUAL                       (4)
```

Yet, if you would have decided to build your own solution from the beginning, the costs would be

```
ALTERNATIVE_COSTS = BUILD_DIRECT_ALTERNATIVE
+ (USAGE + BUILD_MAINTENANCE) * n_months                      (5)
```

Given that the usage costs are similar in both scenarios (see Table 9.1), we have

```
ACTUAL_COSTS - ALTERNATIVE_COSTS =
BUY_DIRECT + BUY_MIGRATE_OFF + BUILD_DIRECT_ACTUAL -
(BUILD_DIRECT_ALTERNATIVE + BUILD_MAINTENANCE * n_months)     (6)
```

In most cases, BUY_DIRECT cost is negligible in comparison to everything else,[139] so as a first approximation

```
ACTUAL_COSTS - ALTERNATIVE_COSTS =
BUY_MIGRATE_OFF + BUILD_DIRECT_ACTUAL -
(BUILD_DIRECT_ALTERNATIVE + BUILD_MAINTENANCE * n_months)     (7)
```

Assuming exactly the same solution being built in both scenarios, the build cost would likely be much smaller in the actual scenario:

```
BUILD_DIRECT_ACTUAL < BUILD_DIRECT_ALTERNATIVE                (8)
```

Or, moving everything to the left side:

---

[138] As for the maintenance costs after the N months, let's just ignore them for both scenarios. Assuming they are equal, we may be belittling ALTERNATIVE_COSTS, but that doesn't affect the results of this analysis, and it helps us avoid introducing an extra variable (M months following the N months).

[139] Both integration costs and finances.

```
BUILD_DIRECT_ACTUAL - BUILD_DIRECT_ALTERNATIVE < 0                  (9)
```

The reason for this is that over the last few months, in addition to benefiting from the off-the-shelf solution, you also had the chance to

- Learn more about your actual requirements
- Understand better the domain, the potential solution, and numerous pitfalls

So, even if in both scenarios you end up with building a solution of a similar quality (which is unlikely), the costs in the actual scenario would definitely be smaller. The key question here is whether the left side of `(9)` would still be smaller than zero once we add `(BUY_MIGRATE_OFF - BUILD_MAINTENANCE * n_months)` to it? If that's the case, the alternative scenario of making "the right decision" from the beginning would actually be more expensive than the "unfortunate" situation of a late pivot we've found ourselves in the actual scenario! Then, according to `(7)`:

```
ACTUAL_COSTS - ALTERNATIVE_COSTS =
BUY_MIGRATE_OFF + BUILD_DIRECT_ACTUAL -
(BUILD_DIRECT_ALTERNATIVE + BUILD_MAINTENANCE * n_months) =
(BUILD_DIRECT_ACTUAL - BUILD_DIRECT_ALTERNATIVE) +
(BUY_MIGRATE_OFF - BUILD_MAINTENANCE * n_months) < 0          (10)
```

⇔

```
ACTUAL_COSTS < ALTERNATIVE_COSTS                                (11)
```

That's a very interesting observation: if using an existing solution for some time allows you to save on maintenance costs and learn enough about the domain, potential implementation pitfalls, and your real requirements to compensate for the migration costs, then starting with an existing solution is the most cost-effective option, even if you decide to migrate off it later. And if the migration costs are relatively small, then starting with an existing solution is almost guaranteed to be the best option. That's why understanding the migration costs for the module in question is so incredibly important when making a build-versus-buy decision. Specifically, you want to understand whether the potential migration costs are likely to be covered by the following benefits from using an existing solution for some time:

- No maintenance costs
- Better understanding of
    - Your real requirements
    - Domain
    - Potential implementation challenges

A combination of the two approaches

In some cases, there is also an option of extending an existing solution or combining it with your own. In this case, costs and risks would mostly be a fractional combination of both options. However, the integration costs may also be significant, especially if an existing solution wasn't designed with extensibility in mind.

> If using an existing solution for some time allows you to save on maintenance costs and learn enough about the domain, potential implementation pitfalls, and your real requirements to compensate for the migration costs, then starting with an existing solution is the most cost-effective option, even if you decide to migrate off it later.

9.7. The open source factor

Unfortunately, the vendor lock-in, security, and reliability risks associated with adopting an existing solution may force you to reinvent the wheel. In extreme cases, you may even decide that costs and risks of both the build and buy options outweigh the benefits of having the functionality in question.

However, there is one more important factor that we didn't mention yet: open source. It addresses the biggest fundamental downsides of existing solutions as it provides you with more options to meet your future needs, and to some extent, even mitigates the vendor lock-in problem (see Table 9.4).

	Direct cost	Migration cost	Future costs		Future risks	
			Usage costs	Maintenance costs	Develop-ment risks	Reliabi-lity and security risks
Build your own thing	High?	Not something you would provision for?	New interfaces to understand and remember	Significant?	Buggy?	?
Use an existing open-source solution	Cheap?	?		Low or none	Contribute or make a repo fork	?
Use an existing proprietary solution	Cheap?	?		Low or none	Hard to fix bugs and add new features	?

Note: White means cheap, and dark gray is super expensive. Question marks indicate no/rough estimates. In a book I have to use shades of gray, but normally you would use red/orange/yellow/green instead (and white for N/A and question marks). You can find examples of the original color-coded tables at `https://codehealthguardian.com`.

Table 9.4. Trade-off analysis of building your own solution versus adopting an existing one.

Sometimes you don't even need to fork the whole repository to add a new feature or fix a problem in an open-source dependency. For example, once we had an issue with some of our Facebook API requests not being properly signed, to an extent that Facebook was threatening us it would disable our app in two weeks. It turned out that the problem was in a single line of code in the open-source OAuth library that we used, and to rollout the fix, we didn't even have to fork the whole open-source library or wait for our contribution to be merged in. All we had to do was to "fork"[140] a couple of Facebook-relevant classes, and that's it. That was a fantastic outcome—the open source factor in action.

Even if you don't understand the building blocks of an open-source solution, you may still be able to find somebody who does, or even learn it by yourself. Thus, open source is often a decisive factor in big stake build-versus-buy decisions, making the buy option even more appealing (see Section 9.6).

Summary

In build-versus-buy dilemmas, default to adopting an existing solution, as reinventing the wheel is rarely a better option. It may be if the implementation is trivial or if existing solutions are too expensive, result in a vendor lock-in, or introduce security or reliability risks that are too high. Worth mentioning, we tend to underestimate the costs and risks associated with *our own* solutions.

Widely known mature modules have a broad range of benefits:

- They are usually better documented, supported, and have fewer bugs.
- They almost certainly won't suddenly cease to exist.
- They often guarantee backward-compatibility.
- They are more likely to have the features you will want in the future.

---

[140] Copy-paste them into our codebase.

However, if you're unlucky enough to ever want something that's not there, the module being widely known and stable only increases the risk of your feature requests being ignored. So prefer open-source modules as they provide you with more options to meet your future needs and even mitigate the vendor lock-in problem to some extent.

Finally, when facing a build-versus-buy dilemma, don't rely on general recommendations—do your own trade-off analysis to make a decision. The devil is in the details.

Chapter 10

CODE REVIEWS

It is extremely rare that the first draft of any code change is good enough. A healthy change is typically the result of multiple iterations of review and improvements. It's a step-by-step process in which each subsequent step provides the necessary insights for the following one. Yet, it's not about iterations only—a fresh perspective is another essential ingredient, and the easiest way to get it is to ask somebody else to take a look.[141]

In this final chapter, we'll discuss such collaborative code reviews. Yes, it is important for authors to review and iterate on their code themselves, but not getting any external opinion is hardly ever the best approach.[142] And this is not only so because of the reviewer's fresh perspective, but also because of their expertise.

In the coming years of AI dominance, it is our expertise that will empower us to outperform AI reviewers. A scientific approach, combined with our creativity, experience, and complex reasoning capabilities will likely help us remain the best candidates for the old-new role of a Code Health Guardian for the next few years.

---

[141] If you want to get a fresh perspective by yourself, then taking a break and getting back to your code the next week might be the best idea.

[142] In the 1980s, Donald Knuth famously designed and implemented TeX all by himself (maintaining an active feedback loop with the TeX community). Yet it's likely that even he himself would be more productive with the help of the formal code reviews as we know them today.

10.1. The four goals of code reviews

A product is normally the reason we write code, and as demonstrated in Chapter 1, the health of a codebase affects the product immensely. So it's no surprise that the primary goal of code reviews is to *ensure there is no code health regression*. To some extent, the author of the change, especially if senior, can verify it themselves, but there is a fundamental problem there—as the author, you know what you meant by your comments, design, naming choice, etc. With that context, it's hard to reason about whether what you actually wrote conveys that meaning. Such curse of knowledge[143] makes a pair of fresh eyes indispensable.

Ensuring code health is obviously the main but not the only goal of code reviews. The second goal, almost as important as the first one, is *providing feedback and knowledge sharing*. There is one common problem with knowledge sharing in our industry—it's the lack of context. It is not that hard to misunderstand what somebody is saying during the meeting or even in a design doc and thus give them feedback that is totally useless. So most of us have learned to be rather skeptical about such high-level feedback and suggestions from our colleagues. But it's a totally different story with code reviews! When reviewing code, you have most of the context necessary right in front of your eyes, and the author of the changes is there, incentivized to clarify things for you. That makes code reviews one of the best opportunities to be *truly* on the same page with our colleagues, which is the most important prerequisite for effective knowledge sharing.

The third goal of code reviews is *building trust*. We don't know yet how important it will be for AI agents, but with humans, building trust is definitely one of the top-level goals. It may sound a bit weird, but code reviews represent one of the most intimate interactions among developers, making it an excellent opportunity to show your colleagues how much you care about them—both long-term (about their growth) and short-term (about their mood today).

---

[143] https://en.m.wikipedia.org/wiki/Curse_of_knowledge

Interestingly, it's not only about being a good human and doing the right thing. It's also a valuable investment in the psychological safety of your team, which is by far the most accurate predictor of its productivity. At Google, research was conducted among more than 200 teams to discover what makes a team productive, an initiative called Project Aristotle. At the beginning, researchers were expecting to find out that the best teams would be those with outstanding individuals put to work together. Instead, the research revealed something surprising:

> *At Google who is on a team matters much less than how team members interact, structure their work, and view their contributions. [...] The number of top performers on a team, and the general intelligence of a team, emerged as a poor indicator of a team's effectiveness. While we found both attributes in some of the best teams, we also found them in some of the weakest. [...] And across all types of teams, from sales to engineering, from San Francisco to Singapore, five dynamics consistently differentiated top-performing teams from those at the bottom* (Natasha Tamiru).[144]

Here they are, in the order of decreasing impact on team productivity (see Figure 10.1):

1. Psychological safety
2. Dependability
3. Structure and clarity
4. Meaning
5. Impact

---

[144] https://www.thinkwithgoogle.com/intl/en-emea/consumer-insights/consumer-trends/five-dynamics-effective-team

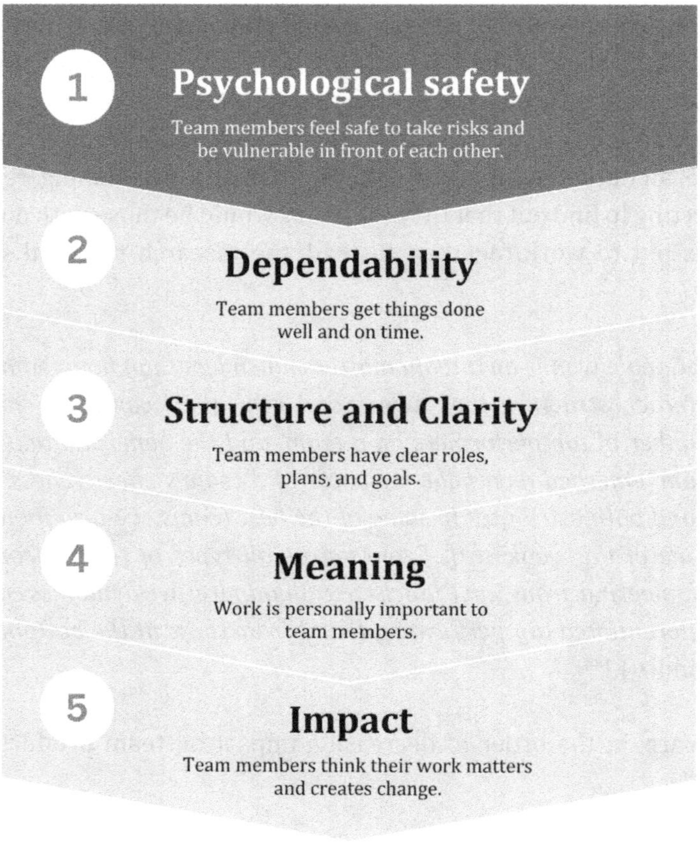

Figure 10.1. What makes Google teams effective.[145]

Tamiru continues,

> *Of the five dynamics, one stands out ahead of the others. Our researchers found that the best teams created a climate of openness where team members admit to their errors and discuss them more often. In other words, they exhibited high levels of psychological safety.*[146]

---

[145] Redesigned `https://rework.withgoogle.com/print/guides/5721312655835136`

[146] `https://www.thinkwithgoogle.com/intl/en-emea/consumer-insights/consumer-trends/five-dynamics-effective-team`

What was even more surprising is that psychological safety is not only the best indicator of a team's effectiveness, but also that the second place is a *distant* second place. There is no single factor that comes close to psychological safety as a predictor of teams' productivity.

There are multiple mechanisms for influencing the level of psychological safety within a team,[147] but code reviews stand out. It's no coincidence Google's internal code review tool is named Critique—when reviewing code, we're faced with our errors more often than anywhere else.[148] Thus, freely discussing our mistakes, modeling being vulnerable, and showing genuine care for our colleagues during the review process is an unprecedented opportunity to gauge the level of psychological safety in the team. Don't ever underestimate the importance of code reviews! They play a crucial role in shaping your team's trust climate.

Last but not least, the code review process shouldn't hinder the *delivery speed* too much. Going back to the first principles, strategic programmers care not only about writing healthy code to deliver high-quality software a year from now but also about delivering it right now (see Section 1.1).

So, here is the final list of the four code review objectives:

- Ensure code health
- Provide feedback and share knowledge
- Build trust
- Deliver the changes fast

---

[147] Yes, the ultimate solution is to hire the right people!
[148] The only competitor is probably post-incident reviews (at Google called *postmortems*). They're out of scope of this book, but you can learn about them from the "Postmortem Culture: Learning from Failure" chapter of *Site Reliability Engineering* [MBJP16]. `https://sre.google/sre-book/postmortem-culture`

There is obviously a tension between some of them, so further down in this chapter, we'll take a look at multiple practical tips and tricks to strike the perfect balance among these competing objectives.

10.2. Should this PR exist?

This chapter is mostly based on internal Google code review best practices, which have been recently made public.[149] However, note that I don't use the Google term Change List (CL) below, but follow the industry standard and use the term Pull Request (PR) instead.

First and foremost, you don't want to waste time reviewing Pull Requests that will never be merged, so you're better off doing a proper screening first. Always start with reading a PR description. There is a number of potential issues with a PR that you can spot without reviewing a single line of code, for example:

- Multiple unrelated changes in a PR => Split it into multiple PRs
- Change that is too large for a high-quality review => Break it down into a series of small changes[150]
- A substantial change that wasn't discussed with the team => Discuss it (ask the author to write a one-pager?)
- It's not a good time to add this functionality => Postpone it
- The change should be made by some other team instead => Hand it over

You won't regret catching such issues early in the process.

---

[149] `https://google.github.io/eng-practices/review`
[150] Don't be shy asking the author to stack multiple changes on top of each other if necessary.
`https://google.github.io/eng-practices/review/developer/small-cls.html#splitting`

How large is "too large"?

Here is an excerpt from one of the most popular internal Google documents (remember, at Google, PRs are called CLs):

"There are no hard and fast rules about how large is "too large." 100 lines is usually a reasonable size for a CL, and 1000 lines is usually too large, but it's up to the judgment of your reviewer. The number of files that a change is spread across also affects its "size." A 200-line change in one file might be okay, but spread across 50 files it would usually be too large.

Keep in mind that although you have been intimately involved with your code from the moment you started to write it, the reviewer often has no context. What seems like an acceptably-sized CL to you might be overwhelming to your reviewer. When in doubt, write CLs that are smaller than you think you need to write. Reviewers rarely complain about getting CLs that are too small."

As you can imagine, at Google, there is tons of technical documentation, but there is one internal document that literally every software engineer is aware of—it's called "Small CLs." That document explains why it is important to keep your CLs small and how to achieve that. Above is an excerpt from that doc, but I strongly recommend reading the original in full, as it has been recently made public: `https://google.github.io/eng-practices/review/developer/small-cls.html`

After reading the description, review the "main" part of the PR—there is typically a file or two with the key changes, all the rest just being repercussions of those. If there are design problems with the main part, don't waste your time reviewing the rest of it, "because if the design problems are significant enough, a lot of the other code under review is going to disappear and not matter anyway."[151]

If when reviewing the code you notice that it's too hard to distinguish between functionality changes and mere refactoring, ask the author to do the

[151] `https://google.github.io/eng-practices/review/reviewer/navigate.html`

refactoring in a separate PR.[152] Solving such puzzles within a PR is not the best idea—even if you *can* solve it, it's counterproductive.

10.3. The six rubrics

What should you look for in a code review? Let's answer this question with the six code review rubrics. Rubrics work great for a variety of our routines: from evaluating a technical interview with a potential hire to writing design docs. They also work great in code reviews—it's very helpful having a set of topics to think about and questions to give an answer to during a PR review.

Test coverage: Are all the necessary tests there?[153] You might want to take a look at the tests early in the process. If done properly, they'll give you an idea of what the change is supposed to be doing. If some tests are missing, you can ask for them and put the review on hold until that's done. Chances are the added tests would catch a bug or two, thus saving you from reviewing the PR twice. Tests are the main contract enforcement mechanism, so they shouldn't be neglected (see Chapter 3).

Correctness: Does it look correct to you? Normally, tests are the best guarantee of software correctness. In code review, we can do our best to compile and run the code in our head; tests actually do compile and run it. Code review catches bugs once; tests catch bugs forever. Yet, some types of problems are nearly impossible to catch with tests. For example, that's the case with race conditions, deadlocks, or infrastructure configuration. Yes, you may rely on assertions in runtime to spot and diagnose such issues (and you *should* have those assertions; see Chapter 3), but it's strongly preferable to catch them earlier. Thankfully, most PRs don't fall into this category, but for those that do, code review becomes crucial for delivering working software. Pay extra

---

[152] `https://google.github.io/eng-practices/review/developer/small-cls.html#refactoring`
[153] And do they pass? In case this check is not automated.

attention to the code that's not properly covered by tests as you are the last line of defense.

Complexity: Is it the simplest change possible? Most of this book is about code health, so I won't repeat myself, except for one piece of advice: pay extra attention to interfaces, especially public ones. As a rule of thumb, the importance of interfaces being true and simple from the start increases alongside their visibility:

```
private method
  < public method
    < public method in a public class
      < company-wide API
        < public API
```

Security: Are there any security issues? Correctness is always important, but it's especially important for access checks to be correct, as a bug in access checks can lead to vulnerabilities. While in practice reviewing security for most PRs is mostly about making sure the code is correct—looking for missing or incorrect access checks—there are also additional things to watch out for, such as hardcoded passwords, API tokens, and cryptographic keys. If you feel that the change you're reviewing has the potential to introduce non-obvious security issues, consider adding a special reviewer who is qualified to assess security.[154]

Documentation: Is there any relevant documentation, and is it (still) accurate enough? While incorrect and incomplete contracts are covered by the complexity rubric, external documentation like wiki or README files is not. It's not fun writing such external documentation, so it makes sense to have a lower bar for its quality. But it doesn't mean it's okay for it to be outdated—incorrect documentation can cause even more confusion than a missing one.[155]

---

[154] Actually, same as with concurrency, accessibility, and internationalization (see Chapter 8).

[155] Struggling with this part? Consider having more design docs and less documentation.

So during the PR review, ask yourself whether any documentation needs to be created, updated, or deleted as a part of the change you're looking at.[156]

Consistency: Does the code follow your style guide, and does the solution fit well into the overall architecture? We've already discussed it in Chapter 2, so just like with the complexity rubric, I won't repeat myself. Code style consistency should mostly be enforced by continuous integration (CI), but it's your responsibility as a reviewer to make sure that the solution fits well into the overall architecture of the system.

Code quality vs. code style

I usually tell my students that if they confuse the terms code quality and code style during the exam, they will get an F straightaway. There are hardly two concepts that are more different from each other. The whole language style guide is typically several pages long. It can be automated by your code formatter and enforced by continuous integration (CI). But most importantly, the only thing we care about with code style is its consistent application. The specifics of any particular style guide typically just do not matter! On the contrary, code quality is extremely important, highly nuanced, and hard to formalize. You can write hundreds of pages on this topic and just scratch the surface. Healthy code is a synonym for a high-quality code, and this book is definitely just an intro into the topic. It requires contextualization, your own practice, and experiments to internalize its content. As a code reviewer, you're responsible for ensuring code health, but it's CI that is mostly responsible for the code style compliance.

---

[156] Google solved it elegantly with g3doc, which is essentially a wiki that can be updated in the same PR together with the code changes.

Here is the final list of the six rubrics to keep in mind when reviewing a PR:

- **Test coverage:** Are all the necessary tests there?
- **Correctness:** Does it look correct to you?
- **Complexity:** Is it the simplest change possible?
- **Security:** Are there any security issues?
- **Documentation:** Is there any relevant documentation, and is it (still) accurate enough?
- **Consistency:** Does the code follow your style guide, and does the solution fit well into the overall architecture?

For each of the six rubrics, it's important to be pragmatic, not pedantic. The golden rule is the following: as long as PR improves the overall health of your codebase, it's okay to approve and merge it.

> As long as PR improves the overall health of your codebase, it's okay to approve and merge it.

10.4. Practical tips and tricks

Now that we've talked about the goals of code reviews and what to pay attention to, let's go over some practical tips and tricks. Honestly, that's my favorite part, because finding the right balance among the following competing objectives of code reviews is truly challenging:

- Ensure code health
- Provide feedback and share knowledge
- Build trust
- Deliver the changes fast

1. Review every line. First things first. To make sure the code under review is healthy, you need to read and understand *every line* of it. The only exception may be auto-generated code that is based on some other code—typically, you would just review the original and trust the code generation tool.

2. Aim to review within one business day. Unless the author of the PR told you otherwise, it's safe to assume that a pending review is blocking them in some way. They may be simply waiting for you to deliver the change, or they may be actively writing more code on top of this PR. Untimely review is problematic in both cases. Interestingly, "most complaints about the code review process are actually resolved by making the process faster."[157] At Google, the expectation is that you "should do a code review shortly after it comes in. One business day is the maximum time it should take to respond."[158]

But to be so responsive, first you somehow need to know that somebody is waiting for your review. There are multiple ways of achieving that: you might have a chat bot constantly reminding you of pending review requests, or there may be a culture of authors messaging the reviewers asking for a faster review. Both of these are basically push notifications, which are probably not the best for your mental health.[159] Another alternative is the pull approach, that is, to regularly check the list of the reviews assigned to you. However, at Google, we found a solution mixing the pull and push approaches to be the most effective and the least distracting—it's a browser plugin with an icon that turns red if somebody is waiting for your review.[160] So whenever you go to Chrome, it would catch your eye if there are any pending review requests, and then you can take a look at your earliest convenience.

---

157 `https://google.github.io/eng-practices/review/reviewer/speed.html`
158 `https://google.github.io/eng-practices/review/reviewer/speed.html#fast`
159 `https://santamaria.wa.edu.au/notifications-and-our-brains/`
160 Tired of the annoying Slack bot, I recently built a GitHub version of it: `https://chromewebstore.google.com/detail/my-turn-github-pr-review/fmcpmjmkhfmljleacbccojadaodihjim` ("My Turn—GitHub PR Review"). Apologies, it's my first TypeScript code.

Doing a first round of review within one business day doesn't necessarily mean it will be merged the same day, but that's fine as "it's even more important for the individual responses to come quickly than it is for the whole process to happen rapidly."[161] If for some reason you can't take a look at the PR within a reasonable period of time, let the author know, and ideally propose an alternative reviewer.

3. Aim to approve quickly. If there are no serious issues with the PR, consider giving an approval straightaway. You can still make comments explaining your minor concerns that need to be addressed by the author prior to merging the PR, but approve and trust them to do that. This way, they won't need to wait for another round of review from your side. That's a powerful trick that not only helps you deliver faster but also builds trust.

4. Label nonsevere comments. Unless you explicitly mark a comment as a minor issue, the author will likely assume you're considering it a serious one. So, if you want to provide feedback effectively and build up trust, not tension, you need to label minor problems somehow. At Google, we widely used the following notation:

- "[Minor] Your comment here" or "nit: Your comment here"—you don't think it's something serious, but you expect the author to fix it prior to merging the PR.
- "[Optional] Your comment here" or "FYI: Your comment here"—you think it might be better to do things differently, but it's up to the author whether to change it or not.

A habit of leaving optional comments usually helps with knowledge sharing and growth. However, leaving too many comments may be overwhelming to the author, and sharing obvious knowledge may even be offensive.

---

[161] `https://google.github.io/eng-practices/review/reviewer/speed.html`

5. Don't be nitpicky or pedantic. Don't try to make the code look like you wrote it yourself. If the author's voice is completely suppressed, something went wrong in the review. And there is no such thing as an ideal PR. Remember the golden rule—as long as the PR improves the overall health of your codebase, it's okay to approve and merge it.

6. Cite your sources. Citing authoritative sources serves two purposes:

- It makes it clear that your comment is not just your opinion.
- It gives the author an option to learn more on the topic.

It's as simple as saying, "Item 16 from Effective Java suggests it's better to avoid such use of inheritance and use composition instead," ideally, with a page number or a link to follow.

7. Assume competence and good will. If there is a problem, it's likely that the author just accidentally missed it. That's actually the main driving force behind code reviews—we all have blind spots and need another pair of eyes with a fresh perspective. But if the author doesn't know something that you do, it's a great opportunity to share that knowledge and help them grow.

8. Provide guidance. It's not your responsibility to write code for the author. Typically, explaining what you expect should be enough (maybe with a code snippet as an illustration). However, there are exceptions when spiking an alternative PR is a faster and more productive approach. For example, that's the case if the author is trying to use the brand-new interfaces that you've recently written, and there are not enough usage examples in the codebase yet. Then, it might be a good idea for you to understand their use case and spike it properly for them.

9. Leave positive comments. Unfortunately, it's not that common for reviewers to leave comments such as "Wow, that's cool!"; "Thanks for refactoring this!"; "Good stuff!"; "Thank you for cleaning this up!"; or "Thanks for a quick turnaround!" That's a shame, because positive comments not only make your colleagues happy and build trust, but also reinforce good practices.

10. Choose your battles wisely. Some code reviews result in conflicts—even smart and intelligent people disagree sometimes, and that's okay. The first thing to do is to try to reach a consensus, but if you can't, stop leaving comments and discuss it in person. If it gets hot, code review comments, same as emails, are the worst form of communication. Be strict on big points, and don't buy "I'll fix it later" promises—"later" usually never happens. At the same time, be pragmatic about it, and don't waste your time discussing minor issues. You can find more information on the topic in the "Handling pushback in code reviews" section of *Google Engineering Practices Documentation.*[162]

Summary

Code reviews are the primary means of keeping your codebase healthy, but that's not their only goal: they're also one of the best opportunities to be on the same page with each other and effectively share knowledge and provide constructive feedback. In addition, they play a crucial role in shaping the trust climate within a team. Last but not least, it's important to deliver quickly and not get stuck in code reviews.

There are multiple practical tips and tricks to strike the perfect balance among these competing objectives:

1. Review every line.
2. Aim to review within one business day.
3. Aim to approve quickly.
4. Label nonsevere comments.
5. Don't be nitpicky or pedantic.
6. Cite your sources.
7. Assume competence and good will.
8. Provide guidance.
9. Leave positive comments.
10. Choose your battles wisely.

---

[162] `https://google.github.io/eng-practices/review/reviewer/pushback.html`

There are six main rubrics to pay attention to in code reviews:

- **Test coverage:** Are all the necessary tests there?
- **Correctness:** Does it look correct to you?
- **Complexity:** Is it the simplest change possible?
- **Security:** Are there any security issues?
- **Documentation:** Is there any relevant documentation, and is it (still) accurate enough?
- **Consistency:** Does the code follow your style guide, and does the solution fit well into the overall architecture?

I believe code reviews will be at the very heart of the old-new role of human software engineers in the upcoming AI era—us still being Code Health Guardians even when most of the code will be written by AI agents.

EPILOGUE

Leo Tolstoy famously wrote, "All happy families are alike; each unhappy family is unhappy in its own way." Over time, I've become increasingly convinced that this holds true for nontrivial codebases as well. Ultimately, it is this realization that led to the idea of writing this book—to publish the story of a healthy codebase. That was in 2022, half a year prior to the release of ChatGPT, which I believe made this topic even more relevant. If there's one thing that's challenging for AI, it's complex reasoning, so managing code complexity (the most intellectually challenging problem in software engineering) seems to be our primary competitive advantage over the AI coding monkeys. At least for some time.

While living in Cupertino, I used to pass by tennis courts seeing the same coach playing with kids. That was more than five years ago, but I can still hear his voice in my head—it was just a single phrase with a strong Mexican accent repeating over and over again: "What you're gonna do? What you're gonna do? What you're gonna do?" He was clearly focused on making his students think positively and proactively about the situations they find themselves in on the court. "Tell me what's your next move!"—he wanted to prevent any situation from paralyzing them. Instead, what he demanded from them was to make a clear decision in a rapidly changing environment. Today, looking at how AI is disrupting our industry, we're all at that decision-making moment, and if anything, we better not be paralyzed by it. As a first step, we need to clearly hear that call to positive thinking: "What you're gonna do?"

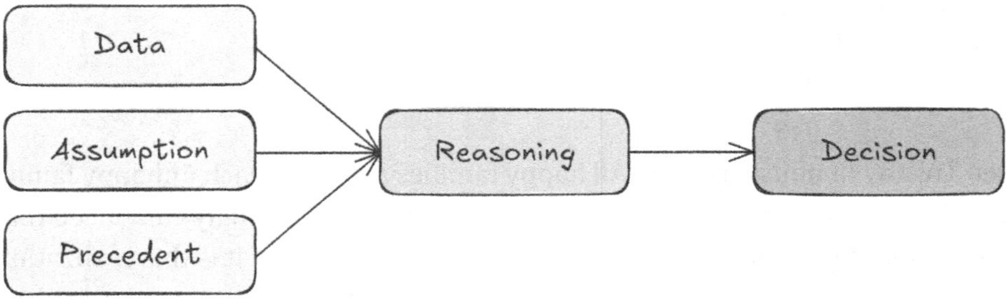

Figure X.1. The four components of real-life decision-making.

Yet, what makes this situation really challenging is the limited number of inputs we have to make our decisions. Google is known to be a data-driven company—internally, it sometimes looks like there is data behind every single decision being made. Yet, in reality, for most of them, there are four components involved (see Figure X.1). Here is how *Software Engineering at Google* puts it:

> *Being data-driven is a good start, but in reality, most decisions are based on a mix of data, assumption, precedent, and argument.* [WMW20]

It's simply because you rarely have enough data to make a truly data-driven decision. Thinking of what we should do to prepare for the upcoming AI era, it's even worse—it's not only the lack of data, but it's also the lack of any reasonable precedents in the past. People compare it to the invention of the wheel, electricity, computer, but even being a strong believer in the value of precedents in decision-making,[163] I feel like all these analogies are flawed. When talking about AI and LLMs in particular, we're talking about a totally new category—a new type of breakthrough that we've never seen before. So if I'm right, assumptions and reasoning are all we have for making our decisions today on how to prepare for the upcoming AI era (see Figure X.2). That looks pretty bad, as it's basically the definition of speculation.

[163] I feel like precedents are generally undervalued, people tending to unproportionally focus on data, assumptions, and reasoning in decision-making. We could have learned more from other people's experience and mistakes.

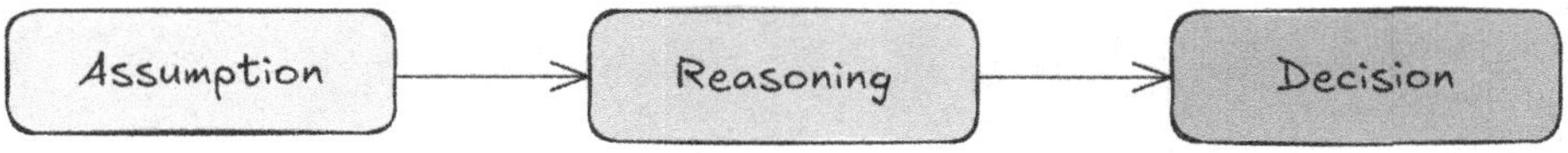

Figure X.2. The two components of decision-making left, when making decisions to prepare ourselves for the upcoming AI era.

How smart will AI get? How fast? If its complex reasoning capabilities tomorrow become better than ours, that would make this book totally redundant.[164] Yet, with an assumption that we'll continue to be smarter than AI for some time, we can still compete with it for the most intellectually challenging parts of our jobs. We, humans, have good chances of remaining the best Code Health Guardians on the market. Maybe for several more years, or who knows, maybe even for several more decades.

If you enjoyed the book, please leave a review!

---

[164] Well, except that AI could be trained on it, but hopefully not.

BIBLIOGRAPHY

[Ashenhurst87] Robert L. Ashenhurst. *ACM Turing Award Lectures: The First Twenty Years, 1966 to 1985.* ACM Press. 1987. ISBN 0201077949.

[Barr18] Adam Barr. *The Problem with Software.* MIT Press Academic. 2018. ISBN 026203851X.

[Bloch08] Joshua Block. *Effective Java.* Pearson Education. 2008. ISBN 0132778041.

[Brooks95] Fred Brooks. *The Mythical Man-Month, Anniversary Edition.* Addison-Wesley Professional. 1995. ISBN 978-0201835953.

[Chesterton09] Gilbert K. Chesterton. *The Thing: Why I Am a Catholic.* Dodo Press. 2009. ISBN 1409931145 (originally published in 1929).

[Evans03] Eric Evans. *Domain-driven Design: Tackling Complexity in the Heart of Software.* Addison-Wesley. 2003. ISBN 0321125215.

[Fowler99] Martin Fowler. *Refactoring: Improving the Design of Existing Code.* Addison-Wesley. 1999. ISBN 0201485672.

[Hoare69] C. A. R. Hoare. *An axiomatic basis for computer programming.* Communications of the ACM. 12: 576–580, 583. 1969.

[HT19] Andy Hunt and Dave Thomas. *The Pragmatic Programmer: Your Journey to Mastery, 20th Anniversary Edition.* Pearson Education. 2019. ISBN 0135956919.

[Kanat-Alexander12] Max Kanat-Alexander. *Code Simplicity: The Fundamentals of Software*. O'Reilly. 2012. ISBN 1449313892.

[Martin08] Robert C. Martin. *Clean Code: A Handbook of Agile Software Craftsmanship*. Pearson Education. 2008. ISBN 0136083250.

[MBJP16] Niall Richard Murphy, Betsy Beyer, Chris Jones, Jennifer Petoff. *Site Reliability Engineering: How Google Runs Production Systems*. O'Reilly Media. 2016. ISBN 1491951176. https://sre.google/sre-book/table-of-contents/

[Meyer88] Bertrand Meyer. *Object-oriented software construction*. Prentice-Hall. 1988. ISBN 0136290493.

[Ousterhout18] John K. Ousterhout. *A Philosophy of Software Design*. Yaknyam Press. 2018. ISBN 1732102201.

[Parnas11] David L. Parnas. *Software Engineering: Multi-person Development of Multi-version Programs.* Lecture Notes in Computer Science. https://www.academia.edu/7337595/Software_Engineering_Multi_person_Development_of_Multi_version_Programs

[Seibel09] Peter Seibel. *Coders at Work: Reflections on the Craft of Programming*. Apress. 2009. ISBN 1430219491.

[WMW20] Titus Winters, Tom Manshreck, Hyrum Wright. *Software Engineering at Google: Lessons Learned from Programming Over Time*. O'Reilly. 2020. ISBN 1492082767. https://abseil.io/resources/swe-book

INDEX

H

I

J

K

L

M

N

O

P

R

S

T

Made in United States
North Haven, CT
10 November 2026

22088899R00117